W0108697

MICHAEL AUFHAUSER

TIERPARADIES
GUT AIDERBICHL

PARADISE FOR
RESCUED ANIMALS

MICHAEL AUFHAUSER

TIERPARADIES
GUT AIDERBICHL

PARADISE FOR
RESCUED ANIMALS

teNeues

Michael Aufhauser & Alexandra

INHALT | CONTENT

GUT AIDERBICHL

EINLEITUNG

Unsere Gut Aiderbichl Höfe nennen viele Menschen „ein Paradies für Tiere". Damit meinen sie nicht nur die gepflegten Anlagen, die romantischen Gebäude und die außergewöhnlich schöne Lage, sondern vor allem die Seele dieser Orte.

Mittlerweile gibt es schon elf Höfe in Österreich und Deutschland, auf denen über 1000 gerettete Tiere glücklich sind. Aber zum Paradies werden die Güter in Henndorf bei Salzburg oder bei Deggendorf in Bayern erst dadurch, dass sich Mensch und Tier nahe kommen und wir sehen können, dass diesen Lebewesen ein Neuanfang geglückt ist. Dass also nichts bleiben muss, wie es ist.

Die traurigen Biografien der Tiere münden auf Gut Aiderbichl in ein Happy End. Wer genau zuhört, erfährt auf Führungen, wie die Tiere in ihre missliche Lage kamen und dass es sich nicht um Einzelfälle handelt, sondern generell etwas schief läuft. Die Situation der Tiere in unserer Zeit hat sehr viel mit den Folgen weltweiter Industrialisierung und der Globalisierung zu tun.

Auf Gut Aiderbichl begegnen Ihnen neben den Rassen, die für die industrielle Landwirtschaft gezüchtet wurden, auch die alten, oft vom Aussterben bedrohten Originale. Von unseren Führern mit Informationen und Wissen versorgt, beginnen viele Besucher darüber nachzudenken, wie man die Beziehung zwischen Mensch und Tier zum Besseren wenden kann. So beginnen Gespräche über rasche Lösungen lang andauernder Miseren.

Nicole Müller & Gustavine

Unsere Tiere haben alle Namen und dazu ein ganz besonderes Versprechen von uns bekommen: Sie dürfen ihr Leben lang, bis zu ihrem natürlichen Tod, bei uns bleiben und müssen nie wieder Angst haben. Wer die Möglichkeit hat, sollte uns besuchen kommen. Dieses Buch wird Sie dabei begleiten. Für diejenigen, die nicht kommen können, soll durch diese Publikation die Möglichkeit entstehen, Gut Aiderbichl trotzdem kennen und verstehen zu lernen. Schließlich geht es nicht nur um Güter, sondern um Ideen.

INTRODUCTION

Many people refer to our Gut Aiderbichl sanctuaries as a "paradise for animals." In doing so, they're not just talking about our neatly kept premises, romantic buildings or the strikingly beautiful scenery surrounding them, but about the soul of these places.

Today, we have eleven sanctuaries in Austria and Germany that have brought happiness to more than 1,000 rescued animals. But what really makes our properties in Henndorf near Salzburg, Austria, or near Deggendorf in Bavaria, such a paradise is that they bring humans and animals close together, showing us that these animals are leading happy new lives and that change is always possible.

Gut Aiderbichl has a way of bringing sad animal biographies to a happy ending. Listen closely to our tour guides and you'll learn what brought these animals to their harrowing ordeals and that their situations, far from being unusual, bespeaks a whole system gone awry. To a large degree, the situation facing animals in our time traces back to the consequences of worldwide industrialization and globalization.

Visiting Gut Aiderbichl, you'll find not only breeds grown for industrial farming, but also the old original breeds, with many on the verge of extinction. Given all the facts and knowledge by our tour guides, many visitors begin to wonder how we can ameliorate the relationship between us and the animals. It encourages discussions about finding swift solutions to on-going misery.

We give all our animals names and, moreover, we make a very special promise to them: They can live at our sanctuaries right up to their natural deaths without ever having to experience fear again. If you ever have the chance, please come visit us and let this book guide you. As for those who can't make the trip out here, well, the reason for this publication is to help you understand and get a feel for our Gut Aiderbichl sanctuaries just the same. After all, it's not just about the product, it's about the idea.

Ein Besuch auf Gut Aiderbichl in Henndorf bei Salzburg

A Visit to Gut Aiderbichl in Henndorf near Salzburg

DIE ANFÄNGE

Hiermit möchte ich, Michael Aufhauser, zu einem Rundgang auf Gut Aiderbichl in Henndorf einladen. Mit dem Bau dieses Gutes wurde im Jahre 2000 begonnen. Die Hauptgebäude entstanden nach den Plänen des Architekten Dr. Michael Ferch. Ursprünglich sollten hier meine privaten Pferde leben. Aber dann kam alles anders.

Wo Gut Aiderbichl steht, verfiel ein verlassenes Bauernhaus. Der Name des Gutes, benannt nach dem Hügel, auf dem es steht, soll keltischen Ursprungs sein. „Ayd" bedeutet im Keltischen „das Feuer". „Bichl" geht auf das althochdeutsche Wort für „Hügel" zurück.

Immer mehr stellte sich heraus, dass dieser Ort inmitten einer traumhaften Heimatfilm-Landschaft sich wunderbar als Begegnungsstätte für Mensch und Tier eignet. So entwickelte sich nach und nach Gut Aiderbichl, wie es heute ist.

Baubeginn im Jahr 2000
Construction start in the year 2000

Mit dem Gut entstand auch eine eigene Philosophie, der sich immer mehr Menschen anschließen. „Die Humanität darf beim Menschen nicht enden!" Dafür steht Gut Aiderbichl und gibt vielen hundert geretteten Tieren die Möglichkeit, den Besuchern zu zeigen, wer sie sind: Teil der Schöpfung, wie wir selbst auch und, verglichen mit uns, die Schwächeren auf dieser Welt, angewiesen auf unsere Rücksicht und Hilfe.

Wenn die Tiere nicht gerade auf ihren Weiden stehen, wo man sie auch besuchen kann, treffen Sie die meisten von ihnen in ihren Stallungen an.

Der alte Ahornbaum steht heute noch. Das Gebäude wurde 1999 abgerissen.
The old maple tree still stands today. The building was razed in 1999

IN THE BEGINNING

Please allow yours truly, Michael Aufhauser, to invite you on a tour of Gut Aiderbichl in Henndorf. Its construction began in the year 2000. The main structures were built on plans submitted by architect Dr. Michael Ferch. The original plan had been for my private horses to live here. But that was before life took a different turn for me.

Gut Aiderbichl stands on what used to be an abandoned, dilapidated farmhouse. The "Gut" part of "Gut Aiderbichl" is actually German for "lot" or "property". "Aiderbichl" is the name of the hill the property stands on and it's said to be of Celtic origin. "Ayd" means "fire" in Celtic. "Bichl" stems from the Old High German word for "hill".

It soon dawned on us that this place and its location in the middle of a gorgeous, *Heimatfilm* landscape—a film genre featuring beautiful countryside, nature scenes and idyllic country living—would make an ideal place for humans and animals to encounter. And so Gut Aiderbichl gradually evolved into what it is today.

The sanctuary also gave rise to a unique philosophy with a growing number of believers. "Humanity isn't just for humans!" This is the credo of Gut Aiderbichl. It gives many hundreds of rescued animals the chance to show visitors that they are as much a part of creation as we humans and that, next to us, they are the meeker creatures of this world, being at the mercy of our compassion and care. And if the animals don't happen to be out on their pastures, where folks can visit them, don't worry: You'll find the majority of them in their stables.

AIDERBICHL – EINE GROSSE TIERFAMILIE, DIE ETWAS BEWEGEN MÖCHTE

Wer Gut Aiderbichl besucht, in Henndorf bei Salzburg oder bei Deggendorf in Bayern, muss nicht überrascht sein, wenn ihm viele freilaufende Tiere begegnen.

Esel, Ziegen, Schafe, vielleicht ein kleiner Stier oder ein keckes Fohlen. Sie sind bei uns zuhause, aber nicht weggesperrt.

Sie durften zu uns kommen, weil sie uns brauchten. Und jetzt brauchen wir sie! Als Diplomaten in eigener Sache. Im gegenseitigen Kennenlernen verstehen viele Menschen besser, weshalb der Umgang mit Tieren etwas über unsere Kulturfähigkeit aussagt und darüber, wie wir mit Schwächeren auf dieser Welt umgehen sollten. Unsere Tierpfleger haben manche unserer Zicklein und Lämmchen mit der Flasche großgezogen und eine enge Verbindung zu ihnen aufgebaut. Dennoch sind diese Tiere ganz natürliche Vertreter ihrer Art geworden. Kaninchen aus Versuchslabors, die Freiheit nicht kannten, haben sich nach kürzester Zeit in ihrem neuen Leben zurechtgefunden. Unsere Haus- und Hofkatzen laufen natürlich frei herum und die geretteten Katzen, die wir vor Gefahren schützen müssen, genießen ihr Leben in der Katzenvilla mit großem Gehege bei Deggendorf: eine Welt der Liebe, Geborgenheit und voll spannender Abenteuer.

Ich selbst leide, wenn ich einen Hund in einem Zwinger sehe. Deshalb leben unsere geretteten Hunde in privater Atmosphäre bei ihrem Pfleger. Wer sie gerne kennenlernen oder Hundepate werden möchte, kann ihnen täglich zwischen 8:00 und 10:00 Uhr auf unserem Gut in Henndorf begegnen. Dort spielen sie vergnügt auf einer eigens für sie eingerichteten Freilaufwiese.

Tristan

Trixie & Lilly

AIDERBICHL—ONE BIG ANIMAL FAMILY
HERE TO MAKE A DIFFERENCE

While visiting the Gut Aiderbichl sanctuaries in Henndorf near Salzburg or in Deggendorf, Bavaria, don't be surprised to see lots of animals out in the open.

Donkeys, goats, sheep, maybe even a little steer or a perky foal—they're all at home here, and we don't lock them up.

We allowed them in, because they needed us. And now we need them! Namely, as diplomats on their own behalf. In getting to know animals, many people develop a better understanding of how our treatment of animals mirrors our level of civilization and how we should treat the more vulnerable species of our planet. Our caretakers have bottle-fed baby goats and baby sheep and developed a close bond to them, which didn't keep these animals in any way from becoming perfectly natural representatives of their species. Test lab rabbits that had never known freedom needed no time at all in

Schnuppi

adjusting to their new lives. Our house and farm cats are welcome to roam freely, and any rescued cats needing protection from outside danger enjoy life in our cathouse on a large reserve out by Deggendorf. Indeed, ours is a world of love, security and full of exciting adventure.

When I see a dog in a kennel, I can't help but feel bad myself. That's why our rescued dogs live in their own private atmosphere with their caretakers. But if you'd like to meet them or if you're thinking about adopting a dog, you can come and see them everyday from 8:00 to 10:00 a.m. at our Henndorf sanctuary where they have their own exercise area for fun and games. Here, they love to play and roam free on a meadow, specially prepared for them.

Rico Fernando & Felix

DAS SCHICKSAL DER PFERDE GEHT ALLE MENSCHEN ETWAS AN

Auf allen Aiderbichler Höfen leben weit mehr als 400 gerettete Pferde, Ponys, Esel und Mulis.

Die Beziehung zwischen Menschen und Pferden in der heutigen Zeit könnte verschiedener nicht sein. Für die einen sind sie Kamerad und Freund, für andere stehen sie im Dienst für Sport und Freizeit. In der Landwirtschaft werden sie noch als Rückepferde zum Abtransport von Holz aus den Wäldern gebraucht oder ansonsten für Festzüge. Sie verrichten Polizeidienste, werden als Reitschulpferde oder als Kutschpferde benutzt, und ein Gros wird im Rennsport eingesetzt. Am Ende ihres Arbeitslebens steht den Pferden in Zentral- und Ost-Europa meistens ein Todestransport in den Süden bevor. Pferdefleisch ist in einigen Ländern Europas, hauptsächlich für die Zubereitung von Wurst (Salami), sehr begehrt. Die meisten Pferdeschlachthöfe befinden sich in Italien. Die Pferde werden dorthin lebendig transportiert und nicht als Fleisch in Kühlzügen, hauptsächlich, weil es kostengünstiger, also profitabler ist. Diese Transporte sind nicht selten qualvoll, und dass man mit schweren Verletzungen bereits vor ihrer Ankunft im Schlachthof rechnet, beweist zum Beispiel ein Krematorium für auf der Fahrt schwer verletzte und verelendete Tiere an der italienischen Grenze in Gorizia.

Nachdenklich macht mich die durchschnittliche Lebenserwartung von sieben Jahren für Pferde in Deutschland, wovon mir Fachleute erzählten. Obwohl doch die Natur eine Lebensdauer von über 30 Jahren für sie vorgesehen hat. Darf nur leben, wer einen Nutzen bringt?

Gigant & Ines

Pferde, die bei uns Aufnahme gefunden haben, dürfen für immer bleiben und werden nicht weitervermittelt. Wir haben ihnen versprochen, dass sie nie wieder Angst, Stress und Druck erfahren. Sie genießen ein tägliches Bewegungsprogramm auf Koppeln oder Weiden, und brauchen bei uns nichts mehr zu leisten. Trotzdem haben Pferde auf Gut Aiderbichl eine ganz wichtige Aufgabe: Mit ihren Lebensgeschichten erreichen sie die Herzen der Menschen. Darunter gibt es immer wieder Pferdebesitzer, Fachleute oder Politiker, die sich, von den Begegnungen beeindruckt, dafür einsetzen, den Pferden dieser Welt bessere Chancen zu geben.

Michael Aufhauser & Toni Bernado

Tiger

THE FUTURE OF OUR HORSES AFFECTS EVERYONE

All Aiderbichl sanctuaries provide a home for at least 400 saved horses, ponies, donkeys, and mules.

The relationship between people and horses today is as complex as ever. Some see horses as companions and friends, while others see them merely as a vehicle for sports and recreation. In agriculture, they're still used as logging horses for hauling timber out of forests, or otherwise as parade horses. They support law enforcement; they're applied as riding-school horses or as carriage horses, while many other horses end up in horseracing. In Central and Eastern Europe, horses having reached the end of their working lives end up mostly on death transports down South. Horsemeat remains very popular in some European countries, mainly in the production of sausage (salami). The majority of horse slaughterhouses operate in Italy. Horses are shipped there in boxcars while still alive rather than as refrigerated meat, mainly because it's cheaper, i.e., more profitable, that way. It's not rare for these transports to cause agony to the horses riding in them. In fact, a crematory at Gorizia, Italy, solely built for animals that were severely injured or killed on these transports proves that their organizers actually anticipate horses to be severely injured before they even arrive at the slaughterhouses.

Pferdemarkt im Herbst

What strikes me about horses in Germany is their average life expectancy of seven years, as I've heard from experts. What about the fact that nature provides them with a life span of at least 30 years? Do their lives depend on their serving some kind of benefit?

The horses at our sanctuaries are free to spend the rest of their lives here without ever being shipped away again. We promise them that they'll never have to experience fear, stress and pressure again. Everyday they enjoy exercise out on our corrals and pastures, and they don't have to serve us in anything. The only task our horses at the Gut Aiderbichl sanctuaries still have is an essential one and that's to touch people's hearts with their life stories. We always see a certain number of horse owners, experts or politicians who are so moved by what they see and hear that they commit to making this world of ours a better place for horses.

Horse market in fall

DER HAUPTSTALL

Mit den Plänen des Hauptstalles haben unser Architekt und ich sehr viel Zeit verbracht. Die Boxen sind 4 x 4 m groß und haben ebenso große Paddocks. Der Ausblick der Pferde ist nicht durch Gitter versperrt. Wir haben darauf Rücksicht genommen, dass sich Pferde einerseits gerne berühren, sich anderseits auch mal zurückziehen möchten. Die Naturholz-Boxen sorgen für eine warme Atmosphäre, sie werden nur in den seltensten Fällen angeknabbert, was für Akzeptanz spricht.

Wir betreten den Hauptstall an der Vorderseite und wenden uns nach rechts. In der ersten Box leben zwei Pferde zusammen, und das hat seinen Grund: Chatleen und Baricello sind Mutter und Sohn. In nächster Nachbarschaft steht die Hessenstute Wanda. Als Reitschulpferd sollte sie wie ein Sportgerät ausgemustert werden. In unmittelbarer Nachbarschaft: Diana und Lotti, Noriker-Stuten. Die Noriker waren die Hauspferderasse der Salzburger Bischöfe. Nach dem Ende des Fürstbischoftums verdanken sie ihre Existenz und Weiterzucht den Bauern im Salzburger Land. Nero, der Schimmelwallach, sollte mit einem Bandscheibenvorfall die qualvolle Tiertransport-Reise in den Tod antreten. Heute lebt er glücklich mit seiner Ponystute Prinzess bei uns. Gleich neben der Türe in der Mitte des Hauptstalles haben seit der Gründung des Gutes Lisa und Ziege Mecki eine Box bezogen. Später kam noch die Ziege Lilly dazu – eine ungewöhnliche WG. Am Ende des Stalles auf der rechten Seite befindet sich eine große Box für Kathi und die beiden Mariandls, auf deren Geschichte ich später eingehen werde.

Den Mariandls gegenüber leben, Box an Box, der Schwarze und Noah. Ex-Showpferde, die mittlerweile unzertrennlich sind.

Natürlich haben auch alle anderen Pferde spannende und lehrreiche Biographien. In der Stallmitte gegenüber des Seiteneinganges seien noch Hilde und Lori erwähnt: die besonders rührende Geschichte eines Haflingerfohlens, das seine Mutter bei der Geburt verloren hatte und in Hilde eine liebevolle Ersatzmutter gefunden hat.

In meinem Buch „Meine schönsten Pferdegeschichten" berichte ich ausführlich über weitere Happy-End-Geschichten von Pferden auf Gut Aiderbichl. Auf einige möchte ich Sie schon an dieser Stelle aufmerksam machen:

Chatleen & Baricello

Diana & Lotti

„Der Schwarze"

Lisa & Mecki

The Main Stable

Our architect and I put a lot of time into the plans for the main stable. Its boxes measure 13 by 13 ft., as do their paddocks. There are no bars to obstruct the horses' views. We kept in mind that horses enjoy physical contact with each other, but that they appreciate privacy too. The natural-wood boxes create a cozy atmosphere and it's exceptionally rare to find bite marks on them, which is a sign of acceptance.

So, let's enter the main stable towards the front and turn right. The first box we see belongs to two horses, and there's a reason for that. You see, Chatleen and Baricello are mother and son. Right next door lives Wanda, a Hessian mare. The riding-school that used to own her had planned to move her out like an old piece of gym equipment. Their other neighbors are Diana and Lotti, two Noric mares. Noric horses used to be the domestic horse breed of the bishops of Salzburg. Since the end of the Prince Episcopate, these horses have owed their ongoing existence and breeding to the farmers of the Salzburg region. Nero, a white gelding, was destined for an agonizing death ride on one of the aforementioned animal transports due to a slipped disc. Today, he leads a happy life at our sanctuary with Prinzess, his pony mare. The box next to the door at the center of the main stable has been home to Lisa and Mecki, the goat, ever since the sanctuary was founded. They were later joined by Lilly, another goat—talk about unusual roommates! At the far end of the stable, to the right, there's a large box for Kathi and the two Mariandls. I'm saving their story for later.

Straight across from the Mariandls, in adjacent boxes, we find the Black Horse and Noah, veteran show horses that have become inseparable.

Of course, all of our other horses have captivating and inspiring biographies as well. And let's not forget Hilde and Lori at the center of the stable, across from the side entrance. Theirs is a particularly touching story of a Hafling foal that lost its mother at birth and found a loving surrogate mom in Hilde.

In my book, "A Happy Ending for Rescued Horses", I will tell you more detailed stories with happy endings for our horses at the Gut Aiderbichl sanctuaries. In fact, allow me to introduce some of them right here:

Wanda

Chatleen und Baricello

Der kleine Baricello kam zu uns als vier Monate altes Schlachtfohlen. Die Trennung von seiner Mutter hatte ihn so traumatisiert, dass es ihm immer schlechter ging. Wir suchten nach seiner Mutter und fanden sie nach einigen Wochen. Chatleen war ein Kutschpferd. Die Wiederbegegnung der Beiden wurde aufgezeichnet, und ein Millionenpublikum konnte vor dem Fernseher Zeuge dieses rührenden Happy Ends werden.

Der Schwarze vom Zirkus

Gemessen am Schicksal vieler Privatpferde, die 23 Stunden am Tag auf ihren Besitzer in einer Gitterbox warten, um dann eine Stunde „bewegt" zu werden, ging es dem Schwarzen beim Zirkus nicht schlecht. 23 Jahre lang nahm er an Tagesabläufen der Zirkusmenschen teil und hatte am Abend seine Auftritte in der Manege. Dass er das Zirkuszelt gegen das Himmelszelt und große Weiden tauschen durfte, verdankt er dem Zirkusdirektor, der sich mit der Bitte an uns gewandt hatte, den Schwarzen bei uns aufzunehmen. Der Zirkushengst ist nun schon 30 Jahre alt und hat sich mit einem Ex-Kollegen aus einer Pleite gegangenen Western-Show befreundet, dem Wallach Noah. Beide können ihre Vergangenheit nicht leugnen, wenn sie erhobenen Hauptes, wie zum Auftritt, auf die Weiden marschieren.

Noah

Kathi und die beiden Mariandls

In einer großen Box des Hauptstalls leben drei silberfarbene Noriker-Stuten. Mutter Mariandl mit ihren mittlerweile ausgewachsenen Töchtern Kathi und Mariandl II. Alles begann damit, dass ich Kathi auf einem Fohlenmarkt als Schlachtfohlen freikaufte und sie auf das Gut brachte. Viele Fohlen leiden so heftig unter der Trennung von ihren Müttern, dass sie lieber sterben wollen, als ohne sie weiterzuleben. So ein Fall war Kathi. Nach einer Woche fanden wir ihre Mutter Mariandl, wurden mit dem Besitzer handelseinig und Kathi bekam ihre Mutter zurück.

Am Tag ihrer Wiederzusammenführung machten wir eine rührende und lehrreiche Erfahrung. Die beiden Pferde erkannten sich, bevor sie sich sehen konnten. Das hatte etwas Mystisches und bestätigte, dass die Wahrnehmungen der Tiere vom Menschen unterschätzt werden. Mutterstute Mariandl war in der Zwischenzeit schon wieder trächtig und so kam auf Gut Aiderbichl Mariandl II zur Welt.

Chatleen and Baricello

Little Baricello came to us as a four-month-old foal bound for a slaughterhouse. He was so traumatized by being taken away from his mother that his health was deteriorating steadily. So we started looking for his mother and located her several weeks later. Chatleen was a carriage horse. The reunion of these two was recorded live, giving millions of viewers the chance to witness this heart-warming happy ending in front of their TV sets.

The Black Circus Horse

Compared to the lot of many privately owned horses that spend 23 hours a day in a steel horse stall waiting for their owners to "exercise" them for maybe an hour, the Black Horse really didn't fare too poorly at the circus. For 23 years, he took part in the daily routine of the circus staff, spending his evenings performing in the ring. Today, he has moved from a circus tent to the great blue tent of the sky, thanks to the circus owner, who contacted us asking if we could give the Black Horse a new home. The Black Horse is now 30 years old and has found a buddy in Noah, a gelding and former show horse himself from a wild-west show that went out of business. Neither one of them can hide their pasts when they march out to the pastures with their heads held high, as if for a performance.

Kathi and the Two Mariandls

A large box in the main stable is home to three silver-colored Noric mares—dam Mariandl and her daughters Kathi and Mariandl II, now grown up. It all began when I bought Kathi at a foal market and took her to our sanctuary to save her from the slaughterhouses. Many foals are so traumatized by being taken from their mothers that they prefer death to life without them. Kathi was one of these. It took us one week to track down Mariandl, her mother, and to work out a deal with her owner, so that Kathi could have her mother back.

Mariandl II, Kathi & Mariandl I

The day they were reunited was a heart-warming and enlightening experience to us. The two horses recognized each other even before they could see each other. There was something mystical to the whole scene. It also proved that humans have a way of underestimating the perceptions of animals. By that time, Mariandl, the mother mare, was already pregnant again, resulting in the birth of Mariandl II at Gut Aiderbichl.

Mücke

Die kleineren Gehege

Über das ganze Gut verteilt leben in kleinen Chalets oder Häuschen im venezianischen Stil haupt-sächlich gerettete Hühner mit ihren Hähnen, sowie Enten oder Gänse und Schwäne. Vielleicht begeg-nen ihnen auf dem Hof einige unserer ausgewachsenen Masthähne. Sie gehören einer neuen Rasse von Hähnen an, die 29 Tage nach ihrer Geburt schlachtreif sind. In der Regel fristen sie ihr Zeitrafferleben in großen monotonen Hallen. Die Natur hat es ursprünglich anders vorgesehen. Einem Hahn blieb früher ein ganzes Jahr Zeit und er durfte wirklich leben, bevor er geschlachtet wurde. Die „patentierten" Hühner aber erhalten nicht selten Medikationen, damit sie schnell wachsen und nicht krank werden. Ihr Gewicht ist so hoch, dass sie die meiste Zeit des Tages im Liegen verbringen. Nur wenige leben länger als ein halbes Jahr. Ihr Herz ist für diese Masse an Fleisch nicht stark genug.

Oliver Gustavine Bernie

Natürlich leben auf Gut Aiderbichl auch Hähne der alten Rassen. Die männlichen Tiere der Lege-hennenrassen werden, kaum dass sie geschlüpft sind, getötet. Sie würden zu langsam wachsen für die Mast und legen natürlich auch keine Eier. Einigen von ihnen konnten wir auf Gut Aiderbichl ein Zuhause geben. Symbolisch wollen wir mit ihrer Anwesenheit auf die tierquälerische Käfig-hühnerhaltung hinweisen.

Alle unsere Enten, Gänse und Schwäne hatten Glück, weil sie Menschen in unsere Obhut gebracht haben. Ihre Vorgeschichten ähneln sich: allesamt haben sie sich in einer ausweglosen Lage befun-den, bevor sie hierher kamen.

Jetzt erinnern sie uns daran, dass wir als Konsumenten wenigstens eines für ihre Art tun können: ein-fach auf Produkte verzichten, die aus der Massentierhaltung kommen – ein erster Schritt für jene, die nicht auf ihren Festtagsbraten verzichten möchten, wäre die Bio-Linie.

Richi

Gustavine

Our Smaller Reserves

All across our sanctuary, we have small *chalets*, i.e., little Venetian-style structures mainly housing rescued hens and their roosters, as well as ducks, geese and swans. At one of our sanctuaries, you might even come across some of our adult roosters bred as fowl. They're part of a new breed of roosters ready for slaughter 29 days after being born. Usually, they spend their fast-forwarded lives in large monotonous halls, contrary to the way originally intended by nature. One rooster actually had a whole year to live out its life before it was slaughtered too. "Patented" fowl, on the other hand, is often kept under medication to make it grow fast and to keep it free of diseases. These chickens weigh so much that they remain stationary for most of the day. Only a few of them live beyond six months, because their hearts can't handle their masses of flesh.

Naturally, Gut Aiderbichl also houses old-breed roosters. Among egg-laying chicken breeds, it's common to kill the males when they've barely hatched. The rationale is that they don't grow fast enough for fowl and, of course, that they don't lay eggs. But we were able to give some of them a home at Gut Aiderbichl. Their presence is our symbolic way of exposing another form of animal abuse—keeping chickens in cramped cages.

All of our ducks, geese and swans were lucky to have humans bring them under our care. Their stories are all similar: they were trapped in a hopeless situation before arriving here.

Today, they remind us that we—their consumers—can do one thing for their species, if nothing else: Simply avoid products from intensive stock rearing. One step in the right direction for those unwilling to go without their holiday feasts would be to try organic.

UNSERE WILDTIERE

Immer wieder erreichen uns Notrufe von Menschen, die sich hilfebedürftiger Wildtiere angenommen haben. Mal sind es Fuchsbabys mit noch geschlossenen Augen, die mit der Flasche großgezogen werden müssen, oder das Bambi hat sich zu einem Rehbock entwickelt und die Retter haben Platzprobleme mit ihrem Schützling.

Dass wir Rotfüchsen ein Zuhause bieten, hat sich herumgesprochen. Bei ihnen handelt es sich eher um scheue Tiere, die nachtaktiv sind und deshalb sehen die Besucher wenig von ihnen. Aber sie haben zutrauliche Verwandte,

Kibib

„Ex-Häftlinge" aus Pelztierfarmen. Ihr Schicksal ist besonders grausam, denn sie leiden für unsere Eitelkeit und Mode. Für einen Pelzmantel müssen über 20 Füchse qualvoll sterben.

Unsere beiden Frettchen Hias und Koarl sind in Henndorf unterwegs und dort finden Sie auch den 30-Ender Rotwildhirsch Burli und seine Begleiterin die Hirschkuh Sophie. Neugierig präsentieren sie sich unseren Besuchern und haben jegliche Scheu vor Menschen verloren. Das gilt auch für die beiden Wildschweine Basti und Reni, die als Frischlinge zu uns gekommen sind. Sehen Sie sich einmal ihren Rücken an. Er ist rund, wie die Natur es vorgesehen hat. Hausschweinen hat man Extra-Rippen ange-züchtet, für noch mehr Koteletts. Sie sind dadurch länglich geworden, was ihnen gesundheitlich oft zu schaffen macht.

Burli

Fuchsgehege / Fox preserve

Hirsch Burli und Sophie

Burli wuchs in einem Rotwildgehege in Oberbayern auf. Er entwickelte von Jahr zu Jahr ein immer schöneres, auffälligeres Geweih. Mit sieben Jahren war er bereits ein würdevoller 30-Ender und Anführer seiner Herde. Sein Besitzer, ein Nachbar und ein kleiner Junge hingen sehr an dem besonders zutraulichen Tier. Wenn Burli gut gelaunt war, durfte der kleine Junge sogar auf seinem Rücken sitzen. Doch musste etwas geschehen, denn es bestand Gefahr, dass Burli seinen eigenen Nachwuchs decken würde. Als man darüber nachdachte, was man tun sollte, meldeten sich die ersten Trophäenjäger. Deren verlockende Angebote für einen Abschuss schlug der Besitzer gerne aus, als sich die Möglichkeit ergab, Burli nach Aiderbichl zu bringen. Gemeinsam mit seiner Hirschkuh Sophie lebt er nun bei uns. Wir selbst haben viel über ihn gelernt. Zum Beispiel, dass er im Frühjahr sein Geweih verliert und es sich in 140 Tagen neu bildet.

Basti und Reni

Der kleine Basti wurde als Frischling zu uns gebracht. Als einziger Überlebender einer Jagd flüchtete er aus einem Wald in ein Schloss und suchte Schutz. Der Schlossbesitzer und seine Haushälterin hatten ein Herz für den Kleinen, zogen ihn auf und brachten ihn dann zu uns. Hier lernte er die ebenfalls verwaiste Reni kennen. Sie sind mittlerweile ein Herz und eine Seele. Und wenn man Glück hat, erlebt man auf Gut Aiderbichl ihren täglichen Badeausflug. Ihre Anwesenheit erinnert uns daran, um welch wunderbare Tiere es sich handelt, und dass wir Menschen die Verantwortung für sie haben.

Eve, Leon & Marc

Basti

Reni & Basti

OUR WILD ANIMALS

We always get desperate calls from people taking in wild animals that are in need. Sometimes it's fox babies, with their eyes still closed, needing to be bottle-fed or cute little "Bambi" is now a full-sized buck driving its saviors out of the room.

We're even known for taking in red foxes. They're pretty cautious animals and become active at night, so visitors don't see much of them. Then again, they do have some sociable relatives, which happen to be former "inmates" of fur farms. Theirs is a particularly cruel fate, because they have to suffer for our vanity and fashion. One fur coat takes 20 or more foxes to die an agonizing death.

Our two ferrets, Hias and Koarl, roam about our Henndorf property, where you can also find Burli, a 30-point red deer stag and Sophie, a hind. She's also Burli's companion. Curious, they present themselves to our visitors, having lost any fear of humans. So have Basti, a wild boar, and Reni, a wild sow, having come to us when they were piglets. Look at their backs for a moment and you'll see that they're round, as intended by

Hias & Koarl

nature. Domestic pigs have been bred to develop extra ribs—for more pork chops. This has made their bodies longer and creates health problems for them.

Burli and Sophie, the Red Deer
Burli grew up on a red deer reserve in Upper Bavaria. Year by year, his antlers grew more beautiful and remarkable. At the age of seven, he was already a handsome 30-point stag and the leader of his herd. His owner, a neighbor and a little boy were very fond of this exceptionally friendly buck. On a good day, Burli even allowed the little boy to sit on his back. The only problem was Burli's potential to create offspring of his own. They were still trying to decide what to do when the first calls from trophy hunters started coming in. They made generous offers for a chance to shoot Burli, but his owner gladly rejected all of them once he had the chance to take Burli to Aiderbichl. Today Burli lives with his hind Sophie at our sanctuary, and we've learned a lot from him. For example, that he loses his antlers in the spring and they grow back after 140 days.

Basti and Reni
Little Basti was brought to us as a piglet. The only survivor of a hunt, he fled from the woods into a castle, seeking protection. Feeling sorry for the little fellow, the castle owner and his housekeeper raised him and then brought him to us, where he met Reni, another orphan. They've been bosom buddies ever since. With any luck, you can watch them go for their daily swim. Their presence exemplifies that these are really precious animals and that we humans bear responsibility for them.

Burli

UNSERE ESEL KÖNNTEN VIEL ÜBER DEN MENSCHEN ERZÄHLEN

Seit mehr als 5 000 Jahren dienen Esel dem Menschen und haben ihn in seiner Entwicklung begleitet. Noch immer werden sie in Ländern der Dritten Welt als Lastenträger eingesetzt oder drehen monoton ihre Runden am Wasserrad. Aber auch in Europa werden sie zum Beispiel zur Ernte von Früchten oder Oliven in unwegsamen, schwer zugänglichen Gebieten herangezogen.

Wenn man auf Esel eingeht, ist man fasziniert, wie intelligent sie sind. Sie sind allerdings nicht ganz so leicht zu handhaben wie ihre Verwandten, die Pferde. Obwohl sie gar nicht störrisch sind, sie denken nur nach. Wenn sie etwas nicht verstehen, halten sie inne und überlegen, ob es sinnvoll ist, was man ihnen abverlangt.

Die meisten der 26 Aiderbichler Esel haben wir entweder auf Schlachtviehmärkten freigekauft oder aus Griechenland geholt. Dort gibt es noch Eselbesitzer, die ihre Tiere, wenn sie ihren Dienst nicht mehr verrichten können, einfach aussetzen und damit zum Hungertod verurteilen.

Englische, griechische und deutsche Tierschützer machen uns immer wieder auf die Verstoßenen aufmerksam. Vielen dieser Esel haben wir bisher ein neues Zuhause auf Gut Aiderbichl geben dürfen. Sie dürfen den ganzen Tag frei herum laufen und machen sich mit ihren Annäherungsversuchen bei unseren Besuchern beliebt. Die meisten sind um die 30 Jahre alt und haben noch gute 20 Jahre vor sich, wenn ihre Gesundheit es gut mit ihnen meint.

Wir züchten nicht. Aber in manchen Fällen retten wir trächtige Tiere und so kam es, dass die Eselin Tina im Jahr 2008 ein kleines Eselbaby mit dem klingenden Namen Donna Anna auf die Welt brachte.

Oh, the Stories Our Donkeys Could Tell about Us!

Donkeys have served mankind for more than 5,000 years while witnessing its development. Today, they're still used for carrying loads or monotonously turning water wheels in third-world countries. Indeed, they're used in Europe, too, for harvesting fruits or olives in impracticable regions hard to access.

Lilly & Chiara

If we take a closer look at donkeys, we can't help but marvel at their intelligence. They may not be quite as easy to handle as their relatives, the horses, but not because they're stubborn—they just use their heads. If they don't understand something, they'll stop to question the wisdom in what they're being told to do.

Most of the 26 donkeys at our Aiderbichl sanctuaries we either bought (freed) from slaughter markets or we imported them from Greece. There you still find donkey owners who will simply set out any animal that can no longer work for them, thereby guaranteeing its death by starvation.

British, Greek and German animal advocates continue to point these poor outcasts out to us and, so far, we've been able to give many of them a new home at Aiderbichl. Some of them stay outdoors all day long, to the delight of our visitors. Most of them are about 30 years of age and still good for another twenty, as long as their health keeps up.

We usually don't see offspring since we're not breeders. However, there are cases when we rescue pregnant animals. In 2008, one of them, our donkey Tina, gave birth to a baby female, which we gave the musical name of Donna Anna.

Donna Anna

DIE GERETTETEN PONYS VON GUT AIDERBICHL

Unter dem Schutz von Gut Aiderbichl stehen insgesamt 69 Ponys. Auf unseren Gütern, in Henndorf und Deggendorf, begegnen Sie einigen von ihnen. Wie sie zu uns kamen, ist schnell erklärt:

In sehr vielen Fällen werden mit der Anschaffung eines Ponys die Sehnsüchte von Kindern befriedigt. Es ist wie mit Schuhen: Die Kinder wachsen heraus und möchten dann entweder ein größeres Pferd oder wenden sich gänzlich einer anderen Welt zu. Deshalb überrascht es nicht, dass die Schlachtpferdemärkte regelrecht überquellen vor Ponys. Dabei könnten sie sogar noch älter als Pferde werden, ein Geschenk der Natur, von dem wenige profitieren dürfen. Mini-Pferde und Ponys gehen uns ausgewachsenen Menschen fast nur bis zur Hüfte. Nicht auf Augenhöhe, lässt sich ein Abschiedsblick, wenn sie der Schlachter holt, leicht vermeiden.

Ernie & Bert

Schecky

Winnetou & Spiky

Alex

THE RESCUED PONIES OF GUT AIDERBICHL

We have a total of 69 ponies under our care at Gut Aiderbichl, some of which you can see at our sites in Henndorf and Deggendorf. The circumstances that brought them to us are easily explained.

Very many cases involve a pony bought to fulfill the wishes of children. It's not unlike buying shoes: The kids grow out of them. Then they either yearn for a bigger horse or their interest turns to entirely different matters. So it comes as no surprise that horsemeat markets are just about overflowing with ponies. The sad part is that they could actually outlive horses, a gift from nature that few benefit from. Mini horses and ponies barely grow beyond the hips of human adults. And not being on the same eye level makes it easy to look the other way when the butcher finally comes for them.

Ernie & Hummel

Cindy & Lena

DIE RINDER VON GUT AIDERBICHL IN HENNDORF

Wenn es um unsere Rinder geht, dann sprechen wir auf Gut Aiderbichl nicht von „Nutztieren", sondern Mitgeschöpfen. Sie werden genauso behandelt wie Hunde, Katzen und Pferde. Und obwohl über 100 Rinder auf unseren Höfen leben, haben sie alle einen Namen, wir kennen ihre Vorgeschichte

Gabi & Heublume

sowie ihre Charaktereigenschaften und wissen, mit wem sie enge Freundschaft pflegen und was sie besonders lieben. Eine Beziehung, wie sie früher für den Bauern und seine Tiere ganz normal war, in einigen Fällen bis heute.

Was Rinder in der heutigen Zeit mitmachen, erfahren wir immer wieder aus den Medien, wenn es zum Beispiel um die langen, qualvollen Tiertransporte geht. Nicht nur in Europa werden die armen Rinder auf lange Reisen geschickt – meistens zum Schlachten. Von Australien oder Brasilien aus verschifft man sie in großen Ozeandampfern lebendig in den Nahen Osten. Aber auch ihre Haltung wird immer anonymer und der modernen Technik angepasst.

Wer sich die Zeit nimmt und unseren Rindern zusieht, lernt viel über ihr soziales Verhalten. Sie nehmen alles um sich herum wahr. Empfinden Glück, Freude und Leid. Trotzdem sind nicht alle Lebewesen gleich und haben ihre Eigenheiten. Rinder zum Beispiel haben keine Gesichtsmuskeln. Das macht es uns Menschen manchmal schwer, sie richtig zu verstehen.

Wenn es Rindern ganz schlecht geht und sie sich in einer verzweifelten Lage befinden, am Schlachthof zum Beispiel, habe ich schon manchmal erlebt, dass ihnen Tränen aus den Augen fließen. Das macht sie uns wieder sehr ähnlich.

Unsere Kühe sind trocken gestellt, das heißt, sie werden nicht gemolken. Nachwuchs bekommen wir in der Regel nur, wenn wir trächtige Kühe aufnehmen. Selbst züchten wir nicht.

Die Haltung von Rindern sollte sich nicht von der Pferdehaltung unterscheiden. Sie brauchen ein tägliches Bewegungsprogramm, Sonne, frische Luft und je nach Jahreszeit abwechslungsreiche Weiden. Natürlich werden sie von unseren besten Pflegern versorgt, medizinisch von Rinderfachärzten betreut, und wenn notwendig kommt ein erfahrener Klauenpfleger zu Besuch.

Sie dürfen bis an ihr natürliches Lebensende bei uns bleiben, werden nicht verkauft oder vermittelt.

Mucky

Zuli

CATTLE AT GUT AIDERBICHL IN HENNDORF

As far as our cattle are concerned, at Gut Aiderbichl, we don't refer to them as "productive livestock" but as fellow creatures. We treat them no differently from dogs, cats and horses. Indeed, even though we have more than 100 cattle at our sanctuaries, they all have individual names, we know their past lives and character traits, who their best buddies are and what their individual affinities are. It's the kind of relationship that used to be a given among farmers and their animals and, in some cases, still is today.

To find out what bovines have to go through these days, just look at what you see in the papers and on TV. You know, news reports of long, torturous animal transports. It's not just Europe sending these poor bovines on their long journeys—usually to slaughterhouses. Large freighters carry them alive to the Middle East from countries like Australia or Brazil. Even cattle farming by itself is turning into an increasingly anonymous and technologically sophisticated industry.

Anybody willing to take time out to observe our bovines can learn a lot about their social behavior. Bovines perceive everything that

Edith & Kärcherl

happens around them. They feel happiness, joy and sorrow too. Still, not all living beings are alike; each has its own unique traits. Bovines don't have facial muscles, for example, making it hard for us humans to understand them sometimes.

When bovines feel terrible and they face a desperate situation, like a slaughterhouse, I've sometimes seen tears flow from their eyes. Doesn't that make them a lot like us?

The cows we keep are dry, meaning we don't milk them. We usually don't see offspring unless we accept pregnant cows. We're not breeders.

Keeping cattle shouldn't be any different from keeping horses. Bovines need daily exercise, sun, fresh air and pastures that change with the seasons. We make sure they're in the hands of our best caretakers and that their medical care comes from top cattle veterinarians. Plus, we rely on the service of an experienced hoof trimmer, as needed.

Our bovines are free to stay with us for as long as they live. We don't sell them or place them anywhere.

Francis

Francis und Gianni

Als ein Viehhändler mit dem 1 400 kg schweren Stier Francis nach Gut Aiderbichl kam, weil er es nicht übers Herz brachte, ihn zu einem Sammelschlachttransport mit Ziel Libanon zu bringen, bauten wir binnen 24 Stunden eine geräumige Hütte. Francis sah sich um, spürte die tiefe Einstreu, entdeckte seinen Eimer voll Äpfel, sowie Heu und frisches Wasser. Da kam er zu mir an den Zaun, leckte mir die Hand und Tränen der Dankbarkeit flossen aus seinen Augen. Aber es sollte noch besser für ihn werden: Er lernte Gianni, einen Angus-Stier, kennen, mit dem er jetzt gemeinsam lebt.

Alexander

Auf der ganzen Welt gibt es nur noch einige hundert Exemplare dieser aussterbenden Rinderrasse: die Pustertaler Sprinzen. Alexander ist ein bisschen zu klein geraten, schied somit als Zuchtstier aus und so kam er zu uns. Auch die Kuh Leslie gehört zu dieser seltenen Rasse und lebt auf dem Gut in Henndorf.

Die Rinderfamilie Harry, Alexandra und Günther

Harry flüchtete beim Verladen in den Wald. Er war gerade von seiner Mutter Alexandra getrennt worden. Sein Besitzer wollte ihn mit Hilfe eines Betäubungsgewehrs zurückbekommen, aber beim Laden löste sich der Pfeil, und er betäubte sich selber. Harry hatte jetzt einen Vorsprung und war drei Wochen lang nicht auffindbar. Als nichts mehr half, brachte man Harrys Mutter in den Wald. Als sie ihn rief, kam er von ganz allein. Alexandra war schon wieder trächtig und bekam Günther, Harrys Bruder. Alle drei sind jetzt erwachsene Rinder und zeigen uns Menschen, was das Wort Familie bedeutet. Sie lieben es, zusammen zu sein.

Francis and Gianni

When a livestock dealer brought a 3,000-lb. steer named Francis to Gut Aiderbichl, because he couldn't find it in his heart to put him on a bulk transport to a Lebanese slaughterhouse, it only took us 24 hours to build a spacious hut for the steer. Francis looked around, felt the deep strew, discovered his bucket filled with apples, plus hay and fresh water. Then he approached me at the fence, licked my hand and tears of appreciation flowed from his eyes. But life was about to get even better for Francis: He made the acquaintance of Gianni, an Angus steer, and they've been living together ever since.

Alexander

The Pustertal Pied Cattle—only a few hundred specimen of this bovine race facing extinction remain worldwide. Alexander was a tad too short to qualify for breeding, which brought him to us. A cow named Leslie also belongs to this rare race. She lives at our Henndorf site too.

Harry, Alexandra and Günther—the Cattle Family

Harry made his getaway into nearby woods during a transport. He'd just been separated from his mother, Alexandra. His owner tried to recapture him by means of a tranquilizer gun, but when he loaded it, the dart came out and he ended up tranquilizing himself. This gave Harry a head start and he disappeared for three weeks. The only option left was to take Harry's mother into the woods. All it took was her call for him and he came back all on his own. Alexandra was already pregnant again and gave birth to Günther, Harry's brother. All three are now fully-grown cattle, showing us humans the true meaning of the word "family" by cherishing every minute together.

Alexander

Michael Aufhauser & Garfield

Garfield

Garfield gehört zur Rasse der West-Highland Rinder und war in seinem früheren Leben Zirkustier. Der Händler erzählte uns, dass er früher in der Manege zur Musik eines Wiener Walzers auftrat. Heute ist er schon sehr betagt. Wir schätzen ihn auf gut 30 Jahre. Hier bei uns braucht er keine Kunststücke vorzuführen, um sich beliebt zu machen. Es genügt, dass er einfach da ist.

Harry, Alexandra & Günther

Mucky

Wenn Mucky sprechen könnte, hätte sie viel zu erzählen. Sie wurde in Anbindehaltung geboren, dann zur Bio-Kuh und nach 15 Kälbern Aiderbichlerin. Sie gehört zur Rasse Grauvieh. Mit über 30 Jahren ist sie das älteste Rind von Gut Aiderbichl und noch immer die Chefin ihrer Herde.

Paul

Als seine letzten Minuten nahten, nahm er allen Mut zusammen. Er ahnte, dass er nichts mehr zu verlieren hat. Stier Paul sprang über die Schlachthofmauer, durchschwamm den Fluss Salzach und versteckte sich in den Flußauen. Als er nach drei Tagen eingefangen wurde, haben die Mitarbeiter von Gut Aiderbichl zusammengelegt und ihn gekauft.

Paul

Garfield

Garfield belongs to the West-Highland breed of cattle and was a circus animal in his old life. His dealer told us that Garfield used to perform to the music of a Vienna waltz back in the ring. Now Garfield has reached a ripe old age. We judge him to be 30 years old, easily. At our place, he doesn't have to show off any tricks to be accepted. We love having him no matter what.

Mucky

If Mucky could talk, she'd have lots of stories to tell. Born in a tie stall, she was an organic cow and delivered 15 calves before becoming a resident of Aiderbichl. Mucky belongs to the gray breed cattle. About 30 years old, she's the oldest bovine at Gut Aiderbichl and still in charge of her herd.

Paul

Knowing his last minutes were near, he gathered all his courage. He knew he had nothing left to lose. So this steer, Paul, jumped the wall of the slaughterhouse, swam across the river Salzach and hid along its banks. Following his capture after three days, the staff of Gut Aiderbichl put together enough money and bought him.

Garfield

Sieglinde und Siegfried

Das kleine Kälbchen Siegfried lag neben seiner Mutter im Rinderstall des Salzburger Viehhofes. Siegfried war gerade einmal ein paar Stunden alt. Ich entdeckte die beiden, als ich wegen der Rettung einer Haflingerstute zum Schlachthof fuhr. Noch am gleichen Tag hielten Sieglinde und Siegfried Einzug auf Gut Aiderbichl.

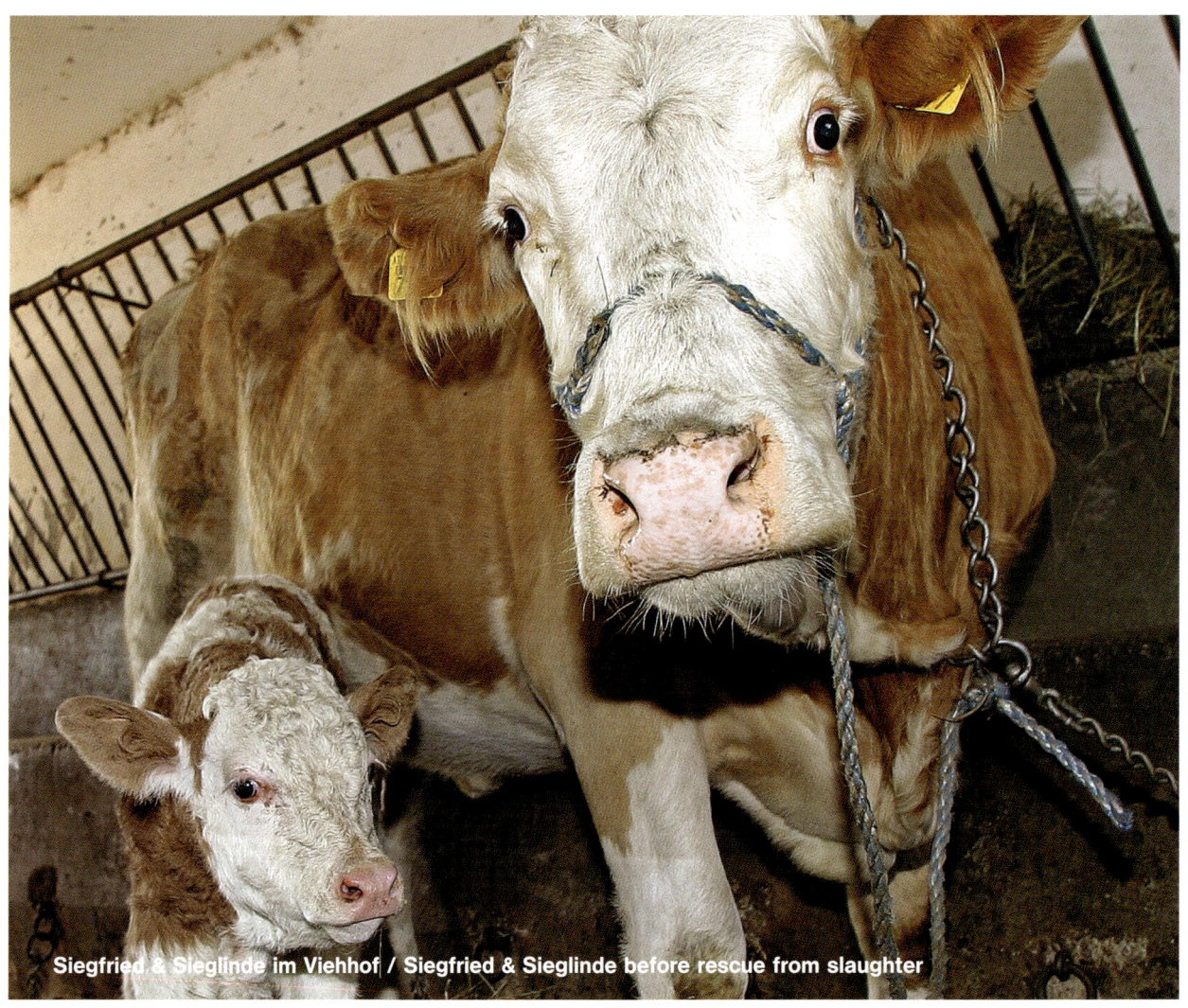

Siegfried & Sieglinde im Viehhof / Siegfried & Sieglinde before rescue from slaughter

Vincero

Als Ralf Schumacher nach Salzburg zog, hat man ihm das Kalb Vincero geschenkt. Sein Name bedeutet auf Italienisch: „ich werde siegen". Ein gutes Omen, denn Schumacher gewann kurz darauf seinen ersten Grand Prix. Vincero durfte leben bleiben und kam zu uns. Er gehört zur Rasse der Pinzgauer.

Insgesamt leben zurzeit auf unserem Gut in Henndorf 37 Rinder. Hier erhalten sie einen guten Überblick, wie viele alte Rassen es gibt. Aber auch eine Vertreterin der modernen Rasse, die Turbokuh Franziska ist ein Teil unserer gemischten Rinderfamilie. Bei uns kann sie entspannt und glücklich leben, ohne Druck, viel Milch geben zu müssen.

Vincero

Sieglinde and Siegfried

Siegfried was a tiny baby-calf lying next to its mother in the cattle barn of the Salzburg stockyard. He was barely a few hours old. I discovered the two of them after arriving at the slaughterhouse to save a Hafling mare. The same day, Sieglinde and Siegfried moved to Gut Aiderbichl.

Vincero

When famous racecar driver Ralf Schumacher moved to Salzburg, the city gave him a little calf named Vincero as a present. His name is Italian for "I shall win". It was a positive sign, because shortly thereafter Ralf Schumacher won his first grand prix race. Vincero was allowed to live and came to us. He belongs to the Pinzgau Cattle.

At this point, we have a total of 37 cattle at our sanctuary in Henndorf. Here, you get a good idea of how many of the old breeds are still around. However, you can find representatives of the modern breeds too, like Franziska, who's a member of our mixed family of cattle as well as a former turbo-breeder for industrial farming. Here with us, she can lead a peaceful and happy life without the constant worry of not producing enough milk.

Tick, Trick, Track & Huberta

Mini- und Hausschweine von Gut Aiderbichl

Schweine werden heutzutage aus Kostengründen meist in großen Beständen gehalten. Oft sind es zigtausende, die in langen Hallen untergebracht werden, in engen Boxen, ohne Stroh und Auslauf. Zu sehen bekommt man sie in der Regel gar nicht mehr.

Und dennoch haben wir die Überzeugung, dass es sehr wichtig ist, gerade unseren Kindern die Möglichkeit zu geben, die Schweine besser kennenzulernen.

Minischweine werden uns meist von überforderten Besitzern abgegeben, die sie als Ferkel zu sich genommen haben und mit einem ausgewachsenen Tier nichts anzufangen wissen. Unsere Hausschweine sind entweder ihrem Todestransport entkommen oder Menschen aufgefallen, die ihnen helfen wollten.

Erfahren Sie auf Gut Aiderbichl etwas über die Hintergründe der Schweinehaltung, der neuesten Zuchten und ob Schweine wirklich die intelligentesten Tiere sind.

Almut & Familie

MINI PIGS AND DOMESTIC PIGS AT GUT AIDERBICHL

The way we see most pigs today is in the form of large stocks on factory farms for cost reasons. Often, many thousands of them are kept in long halls, cramped into tight boxes without any straw or exercise. Usually, we don't even see them outside anymore.

At Gut Aiderbichl, however, we firmly believe it is very important to develop a better understanding of pigs, especially for our children.

Mini pigs are mostly brought to us by overburdened owners who adopted them as piglets without any clue of what to do with those piglets once they grow up. Our domestic pigs either made their escape from some death transport or they caught the attention of people who decided to help them.

At Gut Aiderbichl, you can learn a thing or two about the motives behind pig farming, the latest breeds and whether pigs might actually be the most intelligent of animals.

Almut & family

Ein verhängnisvolles Versprechen

Für die Aufnahmen des Fernsehspielfilms „Das Paradies für Tiere" benötigte die Filmproduktionsfirma Ferkel eines bestimmten Alters. Dafür wurden zwei trächtige Mutterschweine gekauft, und planmäßig kam der Nachwuchs. Der Gedanke, dass man die Filmtiere anschließend wieder holt, war mir unerträglich. Und so versprach ich ihnen, als ich in ihre Box sah, dass sie für immer hier bleiben dürfen. Damals dachte ich nicht daran, dass jedes einzelne zwischen 250 und 300 kg schwer werden würde, und in der Folge ein großer Stall gebaut werden musste, um später über 20 Tieren ein schönes Zuhause bieten zu können.

Obwohl es nicht leicht war und es mich einige schlaflose Nächte kostete, bin ich heute glücklich über meine Entscheidung und kann getrost jede Wiederholung dieses wunderbaren Films sehen. Allerdings möchte ich noch bemerken, dass Schweine mitunter auch streiten. Manchmal bilden sich Feindschaften, und so wurden mittlerweile aus einer großen Gruppe mehrere kleine, was weitere Baumaßnahmen mit sich brachte. Wir wollen ja nicht, dass sie sich beim Streiten verletzen. Ihre Anwesenheit auf dem Gut hat jedenfalls enorme Wirkung. Wer ihnen eine Zeitlang zuschaut, findet sie – auch die erwachsenen Tiere – äußerst sympathisch.

Wenn Schweine ihre Würde verlieren

Es reicht schon, wenn einfallslose Moderatoren oder Journalisten zu einem umgekippten Schweinehänger Bemerkungen machen, wie: „Schweinerei auf der Autobahn". Das soll lustig sein. Wie überhaupt Schweine oft zur Belustigung herhalten müssen. Unser prominentestes Schwein war Biggy, sie wurde anlässlich einer Sylvesterfeier zum Streicheln durch den Saal gereicht. Danach fand sie eine Reinemachefrau verlassen in einer Schüssel, zog sie auf und brachte sie zu uns. Nicht selten werden Schweine verschenkt. Minischweine als Hausgefährten oder Ferkel als Glücksbringer, die dann später auf dem Grill enden. Würdeloser können wir mit Tieren nicht umgehen. Kein Wunder also, dass es zwischenzeitlich mehr als 100 Schweine sind, die unter unserem lebenslangen Schutz stehen.

Biggy

Emilio

Piglet

Huberta

Fateful Promise

In order to shoot a TV documentary about Gut Aiderbichl called "Das Paradies für Tiere," or "The Paradise for Animals," the producing film company needed piglets of a certain age. This led to the purchase of two pregnant sows, and the offspring arrived as planned. Meanwhile, I couldn't stomach the thought of those animals being returned to whatever fate awaited them once they had played their parts. So, seeing them in their boxes, I promised the pigs they could stay here forever. It never occurred to me that every single one of them would grow to weigh between 550 and 660 pounds, and that a huge barn would have to be built at some point to make a cozy home for at least 20 animals.

Looking back on it all, I must admit it was a tall order that cost me plenty of sleep. But I'm still glad I did it and I can watch every rerun of this wonderful documentary with my mind at ease. However, allow me to point out that pigs occasionally get into fights. It can even lead to bad blood, which has already caused the breakup of one large group into several small ones. As a result, we had to add more space, because we don't want any pigs getting hurt once tempers begin to flare. Other than that, their presence at the sanctuary is an immense influence. If you watch them for awhile, you'll actually find them to be very personable—yes, even the adults.

No Respect for Swine

Commenting on a tractor-trailer with a load of swine overturned on the highway, it's bad enough when some witless TV host or reporter says something like "highway turned into pigsty" in an attempt to be funny. That's just one of the countless times that swine are made the butt of jokes! Our most prominent pig is Biggy. She'd been passed around for patting in a ballroom at a New Year's Eve party. After that, a cleaning lady found Biggy forsaken in some bowl, raised her and then brought her to us. It's not rare for pigs to be presented as gifts. Just think of mini pigs presented as house pets or piglets as good luck charms, only to end up on a grill. I can't think of a more disrespectful way to treat animals. So it really shouldn't come as a surprise that we now have more than 100 pigs living under our lifelong protection.

EIN BESUCH AUF GUT AIDERBICHL BAYERN IN DEGGENDORF

A VISIT TO GUT AIDERBICHL BAVARIA IN DEGGENDORF

Gut Aiderbichl Bayern

2006 entstand das Gut Aiderbichl Bayern nahe der Donaustadt Deggendorf. Die Hauptgebäude und ein Dutzend geretteter Pferde, die den Kern des Gutes bilden, sind eine Hinterlassenschaft von Dr. Hatto Egerer, der hier zwölf armen, geretteten Pferden Schutz gab. Inzwischen leben hier 43 Pferde, Esel, Mulis und Ponys, 13 Rinder, 14 Ziegen und Schafe, fünf Schweine, Bronze-Puten, Hähne, Hühner und Kaninchen. Eine Besonderheit ist die 2008 eingerichtete und dem Gut angeschlossene Katzenvilla und eine Großvoliere für Tauben.

Gut Aiderbichl Bavaria

2006 marked the opening of the Gut Aiderbichl Bayern sanctuary near the town of Deggendorf by the Danube River. The main buildings and a dozen of rescued horses forming the core of this sanctuary are the legacy of Dr. Hatto Egerer, who saved twelve unfortunate horses by sheltering them here. Today, it houses 43 horses, donkeys, mules, ponies, 13 cattle, 14 goats and sheep, five pigs, bronze turkey hens, roosters, chickens, and rabbits. Our more distinctive facilities include a cat house, added to the sanctuary in 2008, and a large aviary for pigeons.

Taubenvoliere im Bau / Aviary for pigeons under construction

DER HOFSTALL

Besuchen wir zuerst den Hofstall, der 2006 errichtet wurde. Er beherbergt eine große Tierfamilie:

Das „Nilpferd" aus Franken

Ich staunte nicht schlecht, als mich die Aiderbichlerin Angelika aus Würzburg anrief und mir von einer Kuh erzählte, die im Baggersee und im Main gerne baden geht. Ihr Leben sei jetzt in Gefahr, denn sie hat bei der letzten Besamung nicht aufgenommen und ohne Kalb gibt sie natürlich auch keine Milch. Also machte ich mich auf den Weg nach Franken, lernte den Bauern kennen und sah, wie „Nilpferd", so hatte sie der Bauer getauft, im Main schwamm. Kurz darauf kam sie zu uns und sorgte ordentlich für Wirbel.

„Nilpferd"

Gemeindestier Maxl und seine Amelie

Wir hatten aus der Zeitung erfahren, dass der Gemeindestier von Ohlstadt bei Garmisch zum Metzger sollte, weil er es vorzog, mit den Kühen zu spielen, statt sie zu decken. Erleichtert und kooperativ stimmte die Gemeinde unserem Vorschlag zu, Maxl auf Gut Aiderbichl Bayern einziehen zu lassen. Dort lernte er Amelie kennen. Eine wunderschöne Pinzgauer Kuh. Die beiden wurden das Traumpaar des Gutes bei Deggendorf. Bald schon lebten sie in einer gemeinsamen Box. Bis, ja bis die Schwimmkuh „Nilpferd" eintraf. Während Amelie eine stille, schöne und liebevolle Kuh ist, hat „Nilpferd" das, was Sportlerinnen so attraktiv macht: Spannung, Charisma und Agilität. Beinahe hätte Maxl vor lauter Verliebtheit seine treue Amelie verlassen. Da haben wir uns zu einem kleinen Trick entschlossen. Wir ließen eine Box für „Nilpferd" gleich neben dem Pärchen bauen. Die Geliebte täglich im Blick, erschien das Objekt der Begierde mit der Zeit weniger schillernd, und so war bald wieder alles beim Alten. „Nilpferd" hat in der Zwischenzeit einen anderen, vielleicht noch etwas exotischeren Stier im Auge: Ernst.

Ernst und Conny

Noch lebt der Galloway-Stier Ernst, den sein Besitzer in unsere Obhut gab, als Single. Ein wahres Prachtexemplar seiner Rasse. Auch unsere Kühe finden ihn cool, und zwei Verehrerinnen begleiten ihn auf seinen Weidengängen. Liesl, aus dem Papstort Marktl, verdankt ihr Leben einem Mädchen, das damals acht Jahre alt war und sich für ihre Rettung einsetzte. Conny hat einen berühmten Vater, Stier Francis, der bei uns in Henndorf lebt. Als der Tierhändler das entdeckte, rief er uns an, und Connys Reise zum Schlachthof wurde abgesagt.
Während des Sommers sind Ernst, Conny und Liesl zu einem lustigen Dreierpack geworden. Rührend, wie die beiden Kuhdamen erfinderisch werden, wenn sie sich beliebt machen wollen. Aber auch „Nilpferd" hat Ernst(hafte) Ambitionen.

Max & Amelie

Liesl

Christian Kögl & Ernst

THE FARM STABLE

But let's start with the farm stable, shall we? It was erected in 2006 and shelters a wide family of animals:

"Hippo" of Franconia

I received a pretty amazing phone call from Angelika, an Aiderbichl supporter from the city of Würzburg. She told me about a cow that was partial to swimming in a local artificial lake as well as in the Main River. She went on that this cow's life was in danger, as her last insemination had failed, and without a calf, she wouldn't produce milk. So I made the trip to the German region of Franconia, where I met the respective farmer. That was when I saw "Nilpferd" (German for "hippo"), as the farmer called her, swimming in the Main River. Shortly thereafter, she came to us and caused quite a stir.

Maxl, the Village Steer, and Amelie, his Girl

We read in the paper that the village steer of Ohlstadt near Garmisch-Partenkirchen was destined for the butcher shop, because he preferred playing around with the cows to copulating with them. Relieved, the village community readily agreed to our proposal to take Maxl out to Gut Aiderbichl Bayern. Once there, he met Amelie, a gorgeous Pinzgau cow. They became the dream couple of the Deggendorf sanctuary and before long, they shared a box together. That is, until "Nilpferd", the swimming cow, entered the picture. Although Amelie is a quiet, beautiful and caring cow, "Nilpferd" has all the things that make women athletes so attractive: passion, charisma, and agility. Maxl developed such a crush on her that he almost dumped loyal Amelie. So we came up with a little trick. We had a box built for "Nilpferd" right next to that of our couple. Seeing his flame up close everyday, Maxl found the object of his passion less and less intriguing, and before long, things were back to normal. Meanwhile, "Nilpferd" has had her eyes on another, perhaps somewhat more exotic steer—Ernst.

Ernst and Conny

For now, Galloway steer Ernst remains single, after his owner brought him into our care. He's a fine specimen of his breed. Our cows think he's really cool too, and he already has two lovers accompanying him out to the pastures. Liesl comes from the small village of Marktl, the hometown of the Pope. She owes her life to a girl who was only eight years old when she did everything to save Liesl. Conny has a famous father, Francis, the steer that lives at our Henndorf site. When the animal trader found out about that, he called us and Conny's trip to the slaughterhouse was canceled.

Since this summer, Ernst, Conny and Liesl have become a funny three-some. It's really cute watching these two cows come up with ever new ways of trying to impress him. And don't forget "Nilpferd"—she's apparently got the hots for Ernst too.

Conny

Tragische Rettungen mit Happy End

Kälbchen Lissy wurde mit sechs Beinen geboren. Die Natur hatte ursprünglich Zwillinge vorgesehen, und so kam bereits im Bauch der Mutter die Behinderung zu Stande. Wir ließen Lissy in Wien die beiden überflüssigen Beine wegoperieren. Es geht ihr jetzt genau so gut, wie all den anderen. Paulchen

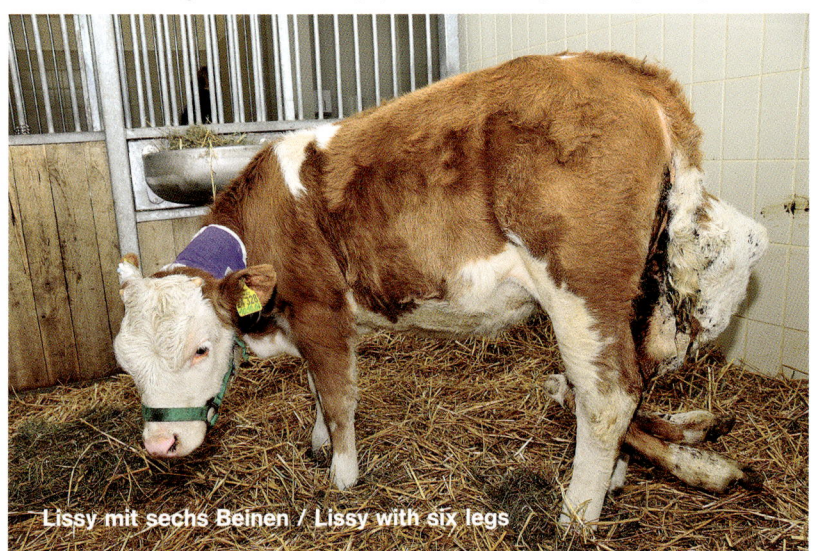

Lissy mit sechs Beinen / Lissy with six legs

wurde seiner sterbenden Mutter aus dem Bauch geschnitten. Sie war gemeinsam mit ihrem Sohn Christian, dem Leitstier, dem Transport zum Schlachter entkommen. Weil Gefahr auf der Autobahn drohte, wurde sie von der Polizei erschossen. Der Leitstier wurde wieder eingefangen und starb kurz darauf an Herzversagen. Nur Paulchen überlebte und erinnert mit seiner Geschichte an die panische Angst, die viele Tiere haben, wenn sie transportiert werden.

Lissy

Alena & Lissy

Dramatic Rescues with Happy Endings

Little calf Lissy was born with six legs. Nature's original plan had been for twins, causing Lissy's deformity while she was still in her mother's womb. So we took Lissy to Vienna, where her two redundant legs were surgically removed. Today, she's doing just as fine as all the others. Paulchen was cut out of his mother's womb as she lay dying. Together with her son Christian, the lead steer, she had escaped on their way to the slaughterhouse. Posing a threat to motorists on the highway, she was shot by police. The lead steer was recaptured and died of heart failure shortly thereafter. Paulchen was the only survivor, his story bearing testimony to the panic experienced by many animals when they are put on transports.

Unsere Schweine, die Glück gehabt haben

Gleich zwei Vertreterinnen der Schweine, die dem Transport entkommen konnten, leben hier im Hofstall. Emma und Lisbeth.

Pünktchen ist mittlerweile ausgewachsen und kann in jungen Jahren auf eine Schauspielkarriere in einer Quizsendung zurückblicken. Maxl sollte als Spanferkel enden.

Nur Pünktchen kannte Stroh, bevor er zu uns kam. Dass man die Boxen mit Stroh ausstattet, sollte eigentlich eine Selbstverständlichkeit sein. Besonders im Winter verbringen die Schweine viel Zeit und Geduld damit, sich Büschel um Büschel weiche Nester zu bauen. Dass die großen Industriebetriebe glauben, Einstreu sei nicht nötig und glatte Rostböden würden genügen, macht das Leben der Schweine in den Mastfarmen oft zur Qual. Ihr Urin ist ammoniakhaltig. Wenn sie gezwungen werden, auf ihrer eigenen Gülle zu leben, leiden sie unendlich. Unsere Pfleger berichten, dass alle Schweine sauber sind. Niemals machen sie dorthin, wo sie fressen oder schlafen. Eigentlich eine einfache Formel, die wir Menschen allen Schweinen gönnen sollten: frische Luft, genügend Platz, Stroh und Wasser. Viel zu viel für sie, meinen Großmäster.

Our Pigs Are the Lucky Ones

Here in the farm stable, we actually have two pigs that escaped being transported—Emma and Lisbeth.

Pünktchen is a fully-grown adult now, who can look back on an acting career in a quiz show when he was still young. Maxl was supposed to end up as a suckling pig.

Only Pünktchen had ever known straw before coming to us. Laying boxes out with straw should be a given. Especially in the winter, pigs invest a lot of time and patience building cozy nests for themselves bundle by bundle. The idea of the big industrial farms that strew is waste and that plain slatted floors are just as good often makes the pigs' lives a living hell. Their urine contains traces of ammonium. Forcing them to live in their own excrements makes their situation unbearable. Our caretakers tell us that all pigs are clean animals that never urinate or defecate anywhere where they feed or sleep. It's really a simple formula we humans should allow every pig to have fresh air, lots of space, straw and water. And a total waste if you ask the industrial farmers.

Das Geheimnis von Esel Pinocio

Auf dem Pferdemarkt wurden hunderte arme Pferde angeboten. Den meisten stand eine qualvolle Reise in den Süden Europas in Akkordschlachthöfe bevor.

Ich kaufte mit meinen Mitarbeitern so viele wie möglich, besonders diejenigen, von denen ich annehmen musste, dass sie zu schwach sind, um den Transport unversehrt zu überstehen. Meine Verhandlungen wurden von dem Geschrei eines penetranten Eselchens begleitet. Egal, wo ich mich in der riesigen Halle befand, er sorgte dafür, dass ich ihn hörte. Eigentlich erhörte. Denn kaum hatte ich auch ihn gekauft, war Pinocio, das Eselchen, still. Auf Gut Aiderbichl schrie er wieder, bis wir ihm Esel Rudi an die Seite stellten. Jetzt glaubten wir, er sei zufrieden. Bis er wieder anfing zu schreien. Weil wir aber gerade keine Eselin zum Dazustellen hatten, brachten wir die Haflingerstute Lizzy zu den beiden in die Box. Seither habe ich ihn nicht mehr schreien gehört.

Pinocio

The Secret of Pinocio, the Donkey

There was a horse fair that traded hundreds of unfortunate horses. The majority of them faced agonizing transports to rate-per-piece slaughterhouses in southern Europe.

Together with my coworkers, I bought as many of them as possible, especially those I had to assume to be too weak to survive the transports unharmed. Suddenly I found my negotiations competing with the mewling of a persistent small donkey. Regardless where I went inside that big hall, that donkey made sure I heard him—or, *listened* to him, I should say. You see, as soon as I bought him too, Pinocio, the small donkey, turned quiet. At Gut Aiderbichl, he mewled again, until we put another donkey, Rudi, by his side, thinking that would pacify him. Then he started to mewl once again. We didn't have a female donkey to put next to him at the time, so we put Lizzy, a Hafling mare, in the box with them. I haven't heard him mewl since.

Rudi & Pinocio

In der gleichen Halle leben auch Ziegen und Schafe. Kommt man ihnen nahe, wie wir und viele Besucher, kann man gar nicht verstehen, dass auf den Menükarten das Wort „Milchlamm" den Appetit mancher Menschen anregt. Eigentlich müsste man doch eher ganz anders reagieren. So wie auf Gut Aiderbichl: sich freuen, dass ein Lamm eine Mama hat und umgekehrt. Als unser Zwillingslamm Jerry an einem Infarkt starb, erlebte ich mit, wie das Muttertier Schnucky und das Brüderchen Tommy tagelang suchten und trauerten. Kismet, werden viele sagen, so geht es nun mal auf dieser Welt. Da hätte ich aber dann noch ganz andere Geschichten auf Lager. Zum Beispiel, dass auf Ocean-Linern, die in Australien auslaufen, bis zu 60.000 Schafe zusammengepfercht werden. Auf der dreiwöchigen Reise in den Tod, zu einem Schlachthof im Nahen Osten zum Beispiel, sterben jährlich 40.000 Schafe nur auf dieser Strecke bereits unterwegs. Vielleicht eine Gnade, denn eine Beschreibung dessen, was sie an ihrem Ankunftsort erwartet, möchte ich Ihnen in diesem Buch ersparen. Da muss man aber doch mal nachdenken: welchen Preis zahlen wir letztlich für einen Lammbraten.

Der absolute Hit für Kinder sind die Ziegen. Es sind sehr viele und sie könnten kaum unterschiedlicher im Charakter sein. Das muss man erleben, erst dann bestätigt sich, dass auch in größeren Gruppen Tiere ihre Individualität behalten, wenn man sie lässt. Aber selbst vor Ziegen macht die Massentierhaltung nicht halt. Umso mehr sind sie auf unsere Sensibilität angewiesen. Ihr Fleisch sollte man jedenfalls nicht ohne Herkunftsnachweis kaufen, soviel steht fest.

Aber auch Pferde stehen im Hofstall. Die Ex-Reitschulpferde Jasmin, Laura, Wolga und Stella. Ein Mädchen, das in einem Reitstall ab und zu mithalf, versetzte sich in das Leben der in die Jahre gekommenen Reitschulstute Jasmin. In ihrem Namen adressierte sie einen Brief an mich und beschrieb ihr Leid. Einige Monate später wurde der Stall veräußert. Wir konnten alle sechs Reitschulpferde kaufen und bei uns unterbringen.

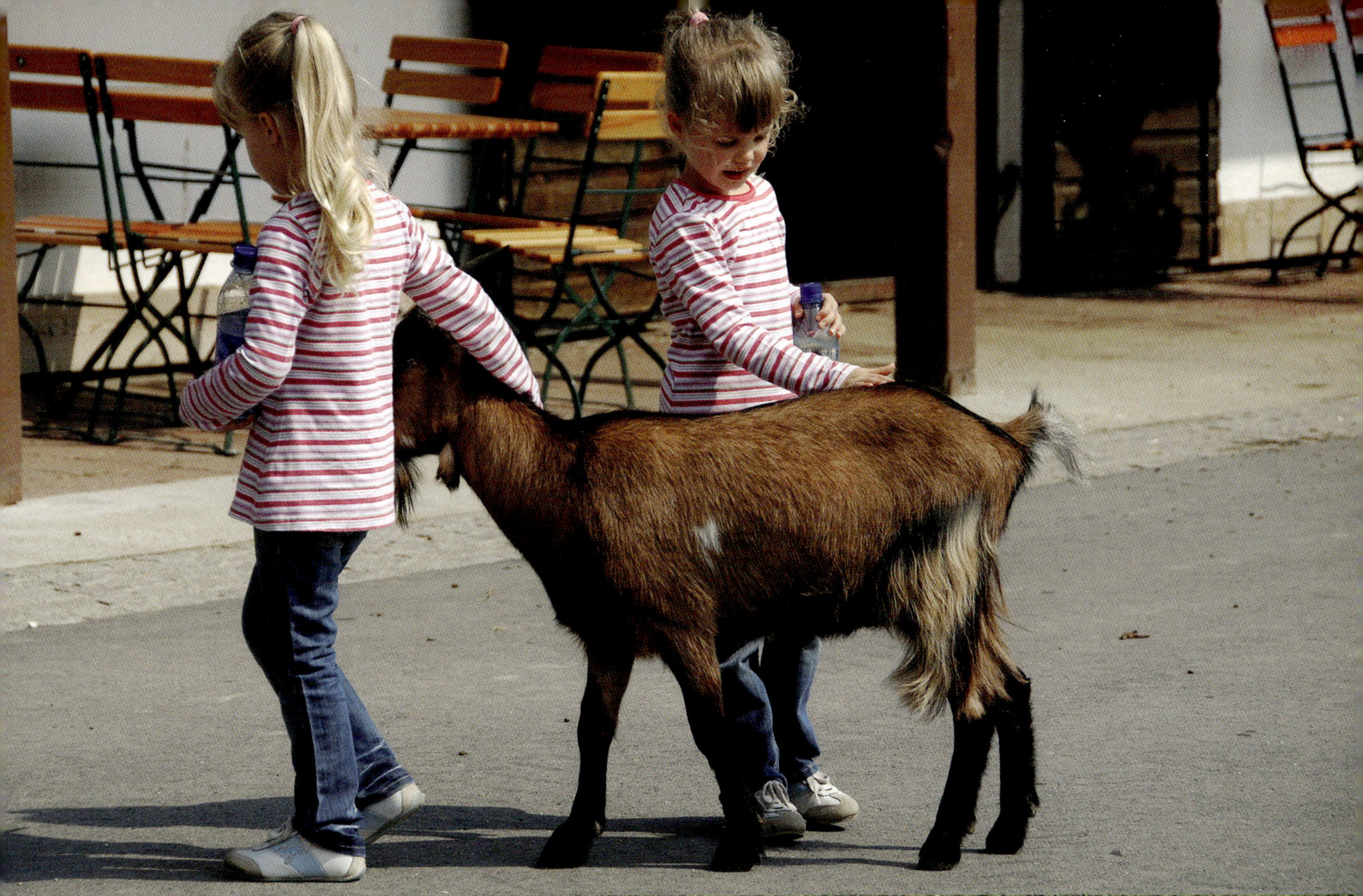

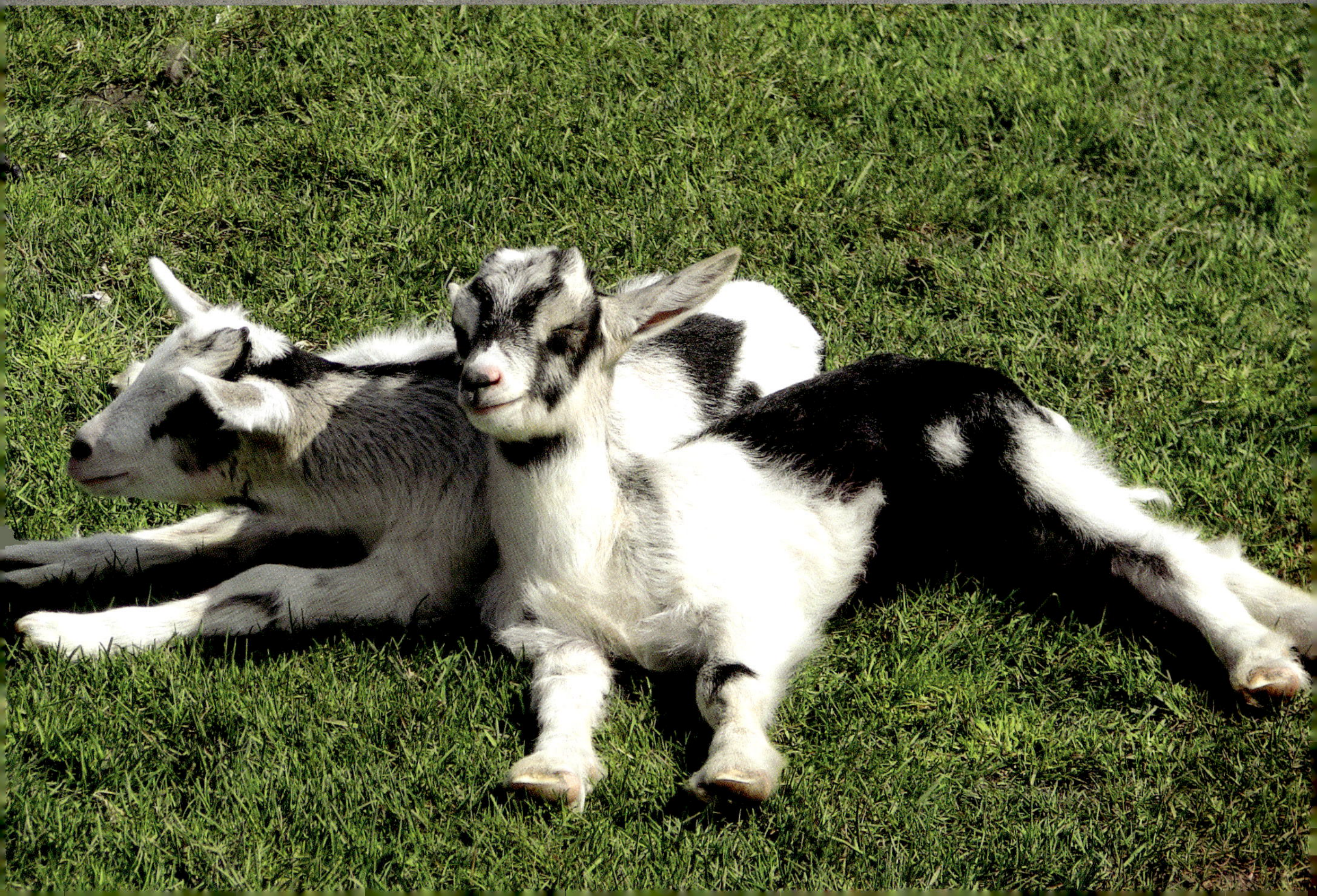

The farm stable also houses goats and sheep. Seeing them up close, the way we and our visitors do, it is mind-boggling how the word "spring lamb" on menus can appeal to the taste of some people out there. Why can't some people just do what we do at Gut Aiderbichl: be glad that a lamb has a mother and vice versa? When our twin lamb, Jerry, died of hypoxia, I saw how Schnucky, his mother, and Tommy, his brother, spent days searching and grieving for him. "Well, that's life," I can

hear some of you say, "that's just the way it is." True, except that I could tell you a whole lot of other stories. For example, ocean liners setting sail from Australia with as many as 60,000 sheep cramped together below deck. It's a three-week death trip to some slaughterhouse in the Middle East, for example, on which 40,000 sheep die every year on this route alone before it even reaches port. Of course, maybe it's an act of God, because, in my book, I want to spare you the description of what awaits the survivors at their destinations. Still, doesn't it make you think about the price we ultimately pay for all those lamb chops?

Among children, our goats are the all-time favorites. We have a whole lot of goats with character traits as unique as they come. You have to see it to believe it: the way animals maintain their own individuality even in large groups if we just let them. Unfortunately, even goats aren't safe from factory farming, which is why they depend all the more on our sensitivity. In any case, I think it's safe to say not to buy their meat without seeing some guarantee of origin.

Of course, we also have horses in our farm stable, namely Jasmin, Laura, Wolga, and Stella, all former riding-school horses. A girl working a part-time job at a riding-stable started to see life from the perspective of aging riding-school horse Jasmin. So the girl wrote me a letter in Jasmin's name, describing her hardship. Several months later, the riding-stable was put up for sale. We were able to buy all six riding-school horses and take them in at our sanctuary.

DER HAUPTSTALL

Dem Hofstall gegenüber liegt ein wunderschöner Innenhof mit einem Brunnen, umgeben vom alten Hauptstall. Im Wohnhaus stehen Ihnen Kaffee und Kuchen und Snacks zur Verfügung, schöne Räume, in denen Sie Aiderbichler Tierfilme sehen können.

Durchfahrt Hauptstall

Bevor man den Stall betritt, gibt es eine große Box in einer ehemaligen Hofdurchfahrt, in der vier Haflinger leben. Drei von ihnen verdanken ihr Leben Christa Clarin, der Witwe des verstorbenen Schauspielers und Tierfreundes Hans Clarin.

Sie hatte bei ihren täglichen Hundespaziergängen in ihrer Nachbarschaft Haflingerfohlen entdeckt und immer wieder mit Karotten und Äpfeln gefüttert. Als sie erfuhr, dass sie, wie die meisten Haflingerfohlen, zum Schlachter gehen sollen, rief sie bei uns an. Jetzt heißt der eine Hans, die andere Christa und die dritte Karin. Und wie so oft auf Gut Aiderbichl konnten wir aus anderen Rettungen noch den kleinen Ottmar dazustellen.

Fohlen, die für die Zucht bestimmt sind, werden von Menschen nach ganz bestimmten Kriterien ausgewählt. Bei Haflingern spielt die Blesse eine große Rolle. Über 90 Prozent der Haflingerhengstfohlen werden deshalb an den Schlachter verkauft, weil die meisten keine perfekte Blesse haben. Unsere geretteten Tiere zeigen aber mit ihrer einnehmenden Art jedem, der hinzuschauen versteht, dass es darauf wirklich nicht ankommt.

Gerettete Schweizer Fohlen / Rescued foals from Switzerland

Christa Clarin

THE MAIN STABLE

Facing the farm stable is a gorgeous inner courtyard with a fountain surrounded by the old main stable. In the farmhouse, you can help yourself to coffee, cake and snacks. There are also nice rooms where you can see animal movies of Aiderbichl.

The Main Stable Passage
Before entering the main stable, you pass a large shed in what used to be the main stable passage. In it, you'll find four Hafling horses. Three of these owe their lives to Mrs. Christa Clarin, widow of the late actor and animal protectionist Hans Clarin.

During her daily routine of walking her dogs, she'd come across some Hafling foals in her neighborhood, and she extended her daily routine to feeding them carrots and apples. Upon hearing that the foals, like most Hafling foals, were scheduled for the slaughterhouse, she called us. Today, the first one's name is Hans, the second one's name is Christa and the third one's name is Karin. And, as it happens so often at Gut Aiderbichl, we were able to place little Ottmar from another rescue in their company.

As in all horse breeds, people choose Hafling foals for very special reasons. The Hafling breed is famous for its blaze. And more than 90 percent of the Hafling stallions are sold to slaughterhouses simply because most of them don't have a perfect blaze. But if you're just willing to take the time, you'll notice a charm in them that makes anyone see life is about a whole lot more than just their blaze.

Im Hauptstall

Hier befinden sich gleich drei Ex-Polizeipferde der Reiterstaffel München. Seit einigen Jahren haben wir eine gute, vertrauensvolle Beziehung zu dieser Dienststelle. Wir sind froh und dankbar darüber, dass wir helfen können und alten ausgedienten Polizeipferden ein lebenswertes Leben bieten können. Immerhin wurden sie während ihres gesamten Lebens als Kameraden von ihren Bereitern gesehen. Ich glaube, auch die Beamten sind erleichtert, dass am Ende dieser Beziehung kein Verrat steht. Quintus ist das bekannteste Ex-Polizeipferd. Er wurde wegen einer Arthrose-Erkrankung dienstuntauglich. Am liebsten verbringt er seine Zeit mit Kronos und Nixon, die früher in seinem Stall standen. So trifft man sich wieder – mit ein bisschen Glück.

Neben den vielen geretteten Pferden leben auch zwei Mulis im Stall, Maryline und Jane. Wir dachten anfangs, sie möchten zusammen bleiben, wurden aber eines Besseren belehrt. Die beiden haben sich Pferde als Partner ausgesucht. Zu wenig wissen wir oft über die Herkunft und über die Geschichte unserer Tiere. Denn meistens finden sie ihren Weg nicht direkt zu uns. Bei jedem Zwischenhändler verlieren sie ein Stück ihrer Identität. Meistens sind sie namenlos, wenn wir sie aufnehmen.

Quintus' letzter Tag bei der Polizei / Quintus' last day as a police horse

Kronos

Kronos

Stefan Gottinger & Quintus

Inside the Main Stable

Here we have no less than three horses that used to be part of the mounted units of the Munich Police Department. We've had a positive and trusting relationship with the Munich Police for a number of years. We are happy and honored to offer them our help by giving retired police horses a life worth living. After all, they've been like companions to their riders all their lives. I believe the officers are glad that this relationship didn't end in betrayal. Quintus is the best-known former police horse. He became unfit for service after developing an arthrosis. His favorite pastime is hanging out with Kronos and Nixon, whom he used to share a stable with. Ah, the joy of seeing old friends again—all it takes is a little luck.

In addition to all the rescued horses, the stable also houses two mules—Maryline and Jane. We used to think they wanted to be together, but we learned better. Both of them have chosen horses as their partners. It goes to show how little we sometimes know about the origins and past lives of our animals. That's because they usually don't come into our ownership directly. First they pass through the hands of various dealers, losing one piece of identity after another. Most of the horses we take in don't even have names.

Jane

Maryline

UNSERE GEFIEDERTEN FREUNDE

Die Bronze-Puten gehören einer aussterbenden Rasse an. Es gibt in ganz Deutschland nur noch 800 ihrer Art. Unsere Putengruppe stand eines Tages am Tor von Gut Aiderbichl. Sie waren einem Hobbyzüchter weggelaufen. Wir einigten uns dann mit ihm, und jetzt dürfen sie für immer bei uns bleiben.

Heino

Ihre Verwandten, die Puten aus den Mastbetrieben wurden so gezüchtet, dass sie viermal so schwer werden können, wie die Bronze-Puten. Insbesondere ihre Brust. Darunter leiden diese Tiere unendlich.

Der kleine Zwerghahn Hansi war von seinen Besitzern in einer Siedlung zurückgelassen worden. Eine tierliebe Dame hat sich seiner angenommen und ihn zu uns gebracht. Damit er nicht einsam ist, hat sie ihm von der gleichen Art noch vier Hennen dazu gestellt.

Auf Gut Aiderbichl Bayern leben außerdem Enten, Gänse, darunter die 37 Jahre alte, blauäugige Irmi, und verschiedene Hühnerrassen.

Die große Tauben-Flugvoliere ist 8 x 18 m groß und beherbergt ganz normale Tauben. Sie sind genauso liebenswerte Tiere, wie alle anderen auch, die bei uns eine Heimat gefunden haben. Indem wir ihre Eier mit Gipseiern austauschen, vermehren sie sich nicht. Wir können immer wieder neuen Tauben helfen.

Hansi

OUR WINGED FRIENDS

Bronze turkey hens are a dying breed. In all of Germany, for example, there are only 800 left. Our turkey hens just showed up one day at the gate of Gut Aiderbichl after running away from an amateur breeder. We then worked out a deal with him and now they can stay with us for good.

Babette

Their relatives, the turkey hens in the factory farming industry, are bred to be four times as heavy as the bronze turkey hens—especially around their breasts. It causes these animals to suffer immensely.

Hansi, the little bantam rooster, had been left behind by his owners in a settlement community. One lady with a love for animals took him in and brought him to us. She even added four hens of the same species in order to keep him company.

Gut Aiderbichl Bayern also is home to ducks, geese, including 37-year-old, blue-eyed Irmi, and various breeds of chicken.

Our big flying cage for pigeons measures 26 by 59 feet and simply houses everyday pigeons. These animals are just as loveable as all the others that have found a home with us. To keep them from proliferating, we replace their eggs with plaster eggs. That way, we keep room for helping new pigeons.

DIE KATZENVILLA

In unserer Katzenvilla leben in der Hauptsache Hauskatzen, deren Besitzer sie nicht mehr halten konnten oder verstorben sind. In einem Zwinger würden sie sich nicht wohlfühlen und deshalb haben wir ihnen ihr Zuhause so eingerichtet, wie sie es ihr Leben lang gewohnt waren. Ihnen fehlt es an nichts, auch nicht an menschlicher Zuneigung, die ihnen von unseren Pflegern täglich entgegengebracht wird. Eine ganz neue Art, verwaisten Samtpfoten über den Verlust ihres Bezugsmenschen und ihrer Heimat hinwegzuhelfen. Unseren Katzen geht es so gut, dass wir sie in der Regel nicht weitervermitteln.

THE CAT HOUSE

Our cat house predominantly houses domestic cats, whose owners were no longer able to keep them or whose owners died. A kennel would make them feel ill at ease, so we set up their new home the way they'd known it all their lives. They have everything they need, including human affection, which they receive from our caretakers every day. It's a totally new way to help orphaned pussycats deal with the loss of their human partners and their homes. Our cats do so well that we usually don't need to find a new place for them.

DIE AUSSENHÖFE

Zusätzlich zu den beiden Besucherhöfen beherbergen noch weitere neun Höfe gerettete Tiere von Gut Aiderbichl. Sie kann man nicht wie die großen Güter in Henndorf und Deggendorf an 365 Tagen im Jahr besuchen. Aber zu besonderen Anlässen ermöglichen wir Förderern von Gut Aiderbichl oder Paten einen individuellen Besuch. Hier seien einige dieser Außenhöfe erwähnt:

Aiderbichl Köllersberger Gut (Henndorf am Wallersee)

2007 kamen über Nacht 21 gerettete Schlachtpferde, die auf einem qualvollen Transport von Rumänien nach Belgien unterwegs waren, unter unseren Schutz. Die Bauernfamilie Köllersberger in Henndorf hatte kurz vorher beschlossen, die Milchviehhaltung aufzugeben. Während die Pferde in Quarantäne waren, bauten wir den Rinderstall um. Aber keine Angst, der Hausstier Florian und vier Kälber fanden auf Gut Aiderbichl ein neues Zuhause. Die restlichen Tiere kamen zu anderen Bauern. Sechs Wochen später waren die Boxen bezugsfertig. Über 30 Pferde erleben jetzt auf insgesamt 17 Hektar Weideland, wie schön das Leben sein kann. Auch Pferde, die früher auf dem Hauptgut lebten, wie Karina und Calvin, Gigant und Angelino, genießen heute die Vorzüge des Köllersberger Gutes. Es gibt eben Tiere, die uns auf ihre Art und Weise zeigen, dass sie ein Leben auf einem ruhigen Hof besser finden.

Dorita & Mary Lou, Köllersberger Gut

OUR SMALLER SANCTUARIES

In addition to our two visiting sanctuaries, we have nine other sanctuaries housing the rescued animals of Gut Aiderbichl. Unlike the two sanctuaries mentioned above, these are not open to visitors year-round. However, on special occasions, we do grant individual visits to sponsors or to promoters of Gut Aiderbichl. Here are some of these smaller sanctuaries:

Aiderbichl Köllersberger Gut (Henndorf/Lake Wallersee)

Back in 2007, we had to put up 21 horses for slaughter over night, after they'd been rescued from a cruel transport from Romania to Belgium. Shortly before that, the Köllersberger farming family in Henndorf had decided to give up on dairy cattle. While the horses were held in quarantine, the Köllersbergers allowed us to rebuild their cattle barn. Of course, we also gave their farm bull, Florian, and four of their calves a new home at Gut Aiderbichl, too. Their remaining animals were taken to other farmers. Six weeks later, the boxes were ready to be occupied. Now we have more than 30 horses enjoying a total of 42 acres of pasture. Life can be so good! Some horses that used to live at one of our main sanctuaries, such as Karina and Calvin, Gigant and Angelino, now find living at our Köllersberger sanctuary a joy, too. Some animals just have their own ways of showing us that they find it preferable to live on a quiet farm.

Köllersberger Gut

Aiderbichl Beruhigungsstall (Henndorf am Wallersee)

Auf Gut Aiderbichl soll jedes Pferd nach seiner eigenen Fasson glücklich werden. Dazu gehört natürlich auch die Option, eher zurückgezogen leben zu können. Im Beruhigungsstall, nahe beim Gut, gibt es einen Freilaufstall und geräumige Boxenhaltung. Die meiste Zeit verbringen die Pferde dort auf Weiden oder Koppeln.

Aiderbichl Beruhigungsstall / Refuge Stable

Manche unserer Pferde haben ein qualvolles Leben hinter sich und furchtbare Angst vor Menschen. Hier erfahren sie durch unsere Pfleger ganz langsam, Schritt für Schritt, dass es Menschen gibt, die es gut mit ihnen meinen.

Aiderbichl Roiderhof (Henndorf am Wallersee)

Während der letzten Jahre fiel uns auf, dass über 20 Pferde und Ponys, darunter auch die Eselin Angelina, sich immer wieder verabredeten und zu einer eingeschworenen Clique wurden. Ihre symbolische Anführerin wurde die Stute Mikina. Wenn wir sie reinholen wollten, um sie wieder in ihre Stallungen auf dem Hauptgut zu bringen, waren sie unglücklich. Nichts wünschten sie sich sehnlicher, als Tag und Nacht zusammen zu bleiben. Im Herbst 2007 war es dann so weit. Jetzt leben sie alle unter einem Dach – auf dem Roiderhof in Henndorf.

Aiderbichl Schroffnergut (Henndorf am Wallersee)

Nicht weit vom Hauptgut in Henndorf gibt es einen malerischen Hof, gleich neben der Kirche St. Birgitta gelegen. Wir konnten unser Glück gar nicht fassen, als wir einen Anruf der Besitzerin Maria

The Aiderbichl Refuge Stable (Henndorf/Lake Wallersee)

At Gut Aiderbichl, we want every horse to discover happiness on their own terms. Naturally, this includes the option of leading a rather private life. Our refuge stable not far away from the sanctuary has an exercise barn and very spacious boxes. Horses also spend most of their time there out on our pastures or paddocks.

Some of the horses we get have seen a lifetime of abuse and are terrified of people. And this is where our caretakers help them understand very gradually and step by step that some people actually have good intentions for them.

Aiderbichl Roiderhof (Henndorf/Lake Wallersee)

We've noticed in recent years how more than 20 of our horses and ponies, including a donkey named Angelina, had a way of constantly getting together and eventually becoming a close-knit group. They've even found a symbolic leader in Mikina, one of the mares. Whenever we'd lead them from the pastures back to their boxes at the main sanctuary, they just didn't like it. There was nothing they wanted more than to hang out together, both night and day. Well, in autumn 2007, their wish came true. Now they all live under one roof at our Roiderhof property in Henndorf.

Aiderbichl Schroffnergut (Henndorf/Lake Wallersee)

Not far from our main sanctuary in Henndorf, there's a picturesque farm right next to the church of St. Birgitta (Bridget). And we just couldn't believe how lucky we were when its owner, Mrs. Maria Gerl, called us on the phone. She was ready to rent out her stable and vast pastures to us on a long-term basis, she told us. It is now inhabited by Pamela and her two offspring, Renzo and Tristan, as well as by 30 other horses. By hopping on our tour ride, which we like to call "Die rasende Biggy," or "Raving Biggy," our visitors can take a tour of the beautiful farm and possibly even spend the night in one of its guest rooms.

Schroffnergut

Gerl erhielten: Wir könnten den Stall und die großen Weidenflächen langfristig pachten, sagte sie uns. Pamela mit ihren beiden Söhnen Renzo und Tristan leben jetzt dort mit über 30 anderen Pferden. Mit unserem Ausflugszug „Die rasende Biggy" kann man den schönen Hof besichtigen und vielleicht auch in einem der Gästezimmer übernachten.

Gut Aiderbichl Köglerhof in Kärnten (Micheldorf bei St.Veit)
Hiltraud und Architekt Andreas Merkl sind Ur-Aiderbichler. Von Anfang an dabei. Im Jahre 2007 schenkte das Ehepaar seinen Hof in Kärnten der gemeinnützigen Stiftung Aiderbichl. Durch seine Ausrichtung wird dieser Hof in der Hauptsache Wildtierfindlingen zu Gute kommen: verletzten Rehen, Füchsen, Wildvögeln, aber auch kleineren Aiderbichler Tieren wie Schafen und Ziegen. Außerdem soll hier eine dauerhafte Ausstellung entstehen, die über alternative Energien aufklärt.

Gut Aiderbichl Kilb (nahe Melk an der Donau)
Am Karfreitag 2008 wandte sich der Bauer Leopold an mich. Er war mit seinen 60 Rindern völlig überfordert, und ihm drohten schlimme behördliche Schritte. Die Rinder zu verkaufen, kam aber für ihn nicht in Frage. Es war ein fast unlösbares Problem, das dennoch wie in einem Märchen endete: Aus dem Hof wurde das Aiderbichl Kilb-Dirndltal. Leopold und die Rinder haben natürlich ein lebenslanges Wohnrecht.

Und dann gibt es noch andere Außenstationen, wie zum Beispiel das Katzenhaus in Fridolfing, die Pferdestation in Tanham oder das Rinderfreigehege in Nesselbach.

Gut Aiderbichl Kilb

Köglerhof

Gut Aiderbichl Köglerhof in Carinthia, Austria (Micheldorf near St.Veit)

Architect Andreas Merkl and his wife, Hiltraud, are original Aiderbichlers. They've been there right from the get-go. In 2007, the couple donated their farm in the province of Carinthia to the not-for-profit Aiderbichl Foundation. The layout of their property will primarily benefit abandoned wild animals: injured deer, foxes, wild birds, as well as more domestic members of the Aiderbichl animal family, such as sheep and goats. There are also plans for a permanent exhibition on alternative energies.

Gut Aiderbichl Kilb (near Melk on the Danube)

On Good Friday, 2008, our good friend, Farmer Leopold, came to talk to me. Hopelessly overburdened by his 60 cattle, he was in hot water with the local authorities. Selling the cattle, though, was out of the question for him. His was an almost insurmountable problem that nevertheless had the ending of a fairytale: His farm became the Aiderbichl sanctuary of the Dirndltal valley. Leopold and his cattle, of course, are entitled to a lifelong residence there.

And let's not forget our other smaller sanctuaries such as our house for cats in Fridolfing, our horse farm in Tanham, and our cattle reserve in Nesselbach.

Gabi Wirths & Lovie, Tanham

Edda & Bunny, Tanham

DAS TEAM VON GUT AIDERBICHL

Es ist ein großes Team aus Helfern, die mit ihrer Leidenschaft den Erfolg von Gut Aiderbichl erst möglich machen. Sie sorgen durch ihr tagtägliches Engagement dafür, dass wir so vielen Tieren ein neues und sicheres Zuhause bieten können. Mit größtem Dank erwähne ich stellvertretend einige der Mitarbeiter, die sich schon länger als fünf Jahre für die Idee von Gut Aiderbichl einsetzen.

Paul Kaiser:

Es sind mittlerweile schon mehr als zehn Jahre gemeinsamer Arbeit, die mich mit Paul Kaiser verbinden. Pauls Leben wird inzwischen von seinen drei wunderbaren Kindern Pia, Paul und Michelle, meinem Patenkind, begleitet. Sein Einsatz galt am Anfang der Gutsverwaltung und der Betreuung der geretteten Tiere. Heute ist er dafür zuständig, dass sich unser Kreis der Aiderbichler nicht nur konstant hält, sondern immer weiter wächst. Unsere Patenschaften sind mittlerweile zu seinem Lebensinhalt geworden.

Dieter Ehrengruber:

Nach seinem Sport- und Geografiestudium arbeitete Dieter Ehrengruber für eine Weltmarke im Sportartikelbereich. Als vor sechs Jahren seine Tochter Laura geboren wurde, entschloss er sich, zu uns zu kommen. Wir verdanken seinem Ideenreichtum und seinem unermüdlichen Einsatz ein gesundes wirtschaftliches Fundament und die beachtliche Größe, die Gut Aiderbichl mittlerweile erreicht hat. Obwohl immer im Einsatz für die Geschäftsführung und die Vergrößerung unserer Standorte, legt er großen Wert darauf, die individuelle Nähe zu den Tieren beizubehalten.

THE GUT AIDERBICHL TEAM

The success of Gut Aiderbichl wouldn't be possible without the passion of a big team of helpers. It is through their daily commitment that we can provide so many animals with a new and safe place to live. It is with utmost gratitude to this whole team that I mention some of its members who have supported the concept of Gut Aiderbichl for more than five years.

Paul Kaiser:

Paul Kaiser and I have been working together for more than ten years. Today, Paul's life also revolves around his three wonderful children Pia, Paul and Michelle, my godchild. Paul first took charge in the management of the sanctuaries and in the treatment of the rescued animals. Today, his responsibility is not just to maintain the number of members in our 'Club Aiderbichl,' but also to keep it growing. Our sponsorship program has become the great mission of his life.

Dieter Ehrengruber:

Having earned his graduate degree in sports and geography, Dieter Ehrengruber first worked in the sports goods line of a global brand. When his daughter Laura was born six years ago, he decided to join us. It is to his abundance of ideas and his untiring commitment that we owe the healthy economic foundation and remarkable size that Gut Aiderbichl enjoys today. As dedicated as he is to the business management and expansion of our sanctuaries, Dieter also cherishes his close individual contact with the animals.

Dieter Ehrengruber

Helmut Schödel:

Gut Aiderbichl wird seit vielen Jahren begleitet von der Arbeit des bekannten Journalisten, Autors und Theaterkritikers Helmut Schödel. Vielleicht war es die innige Beziehung zu seinem Hund Tommy, die ursprünglich den Anstoß dazu gab. Bei seinem ersten Besuch, er schrieb für die Süddeutsche Zeitung einen Artikel über Gut Aiderbichl und seine Philosophie, unternahmen wir einen nächtlichen Spaziergang durch Salzburg. Danach stand für ihn fest, dass er sein Können in den Dienst unserer Sache stellen würde. Er ist seither mein persönlicher Mentor und trägt zur Entwicklung von Gut Aiderbichl wesentlich bei.

Hans Eder:

Hans Eder begleitet mich und unsere Hunde, Katzen, Kaninchen und viele andere Tiere in meinem Privathaus, wie auch Huhn Heidi, tagtäglich mit liebevoller Fürsorge. Er sei symbolisch für all die Menschen genannt, die ganz nah bei den Tieren sind und ohne die es Gut Aiderbichl nicht gäbe.

Hans Wintersteller:

Gut Aiderbichl verzichtet bewusst auf Betitelungen seiner Mitarbeiter. Wer ist also Hans Wintersteller? Er ist Familienvater, Henndorfer, Tierpfleger, Tiertransporteur, Elektriker und Landwirtschaftsexperte in einem. Man könnte ihn auch als Gutsverwalter sehen, aber dann hätte er ja wieder einen Titel und wäre nicht, was er ist: ein Allrounder für die Tiere, Tag und Nacht. Hans Wintersteller ist aber auch ein Teamplayer, der sicherstellt, dass Dr. Hergard Spielvogel und ihr Team von Gutsführern, sowie Anna Pieringer und Maria Kaindl aus Shop und Gastronomie mit ihren Mitarbeitern, unseren Besuchern einen optisch perfekten Ort präsentieren können.

Paul Kaiser

Michael Aufhauser & Helmut Schödel

Hans Eder

Helmut Schödel:

For many years, Gut Aiderbichl has appeared in the work of well-known journalist, author and drama critic Helmut Schödel. Maybe it was his close bond to his dog Tommy that originally motivated him. When he first visited us while working on an article about Gut Aiderbichl and its philosophy for the German daily *Süddeutsche Zeitung*, I invited him out for an evening walk through Salzburg. From that point on, his mind was made up to dedicate his talent to the promotion of our cause. Ever since, he's been my personal mentor as well as a vital contributor to the development of Gut Aiderbichl.

Hans Eder:

Not a day goes by without Hans Eder supporting me and our dogs, cats, rabbits and scores of other animals at my private residence, including Heidi, the hen, with his devoted care. I think of him as the embodiment of all people who relate to animals on a real deep level, and without whom Gut Aiderbichl wouldn't exist.

Hans Wintersteller

Hans Wintersteller:

One thing we never use at Gut Aiderbichl is official job titles. All right—but who's Hans Wintersteller? Well, he's a dad, a native of Henndorf, an animal caretaker, an animal transporter, an electrician and an agricultural expert all rolled into one. I guess you could call him property manager too, but that would mean giving him an official title that doesn't express what he really is—an all-rounder for our animals, both day and night. Moreover, Hans Wintersteller is a team player, who makes certain that Dr. Hergard Spielvogel and her team of sanctuary guides as well as Anna Pieringer and Maria Kaindl with their merchandise and catering staffs keep Gut Aiderbichl neat and presentable to our visitors.

Bernadette Linasi und Christian Dutz:

20 Kilometer entfernt von Gut Aiderbichl in Henndorf läuten die Telefone, wird im Internet geklickt, gebucht und verbucht, werden von Holde Sudenn Bestellungen von Reisegruppen entgegengenommen oder Aufnahmen durch Fernsehteams arrangiert. Und Friedel Grünthal faxt uns die weltweiten Tiernachrichten.

Bis in die Nacht hinein sieht man das Licht brennen, wenn ich z.B. mit Christian am Computer sitze, den ich selbst gar nicht bedienen kann. Hier schlägt ein weiteres Herz von Gut Aiderbichl – im Hintergrund, aber mit genauso starker Vehemenz wie das draußen bei den Tieren. Ohne die Leit- und Schaltzentrale wäre dies alles einfach nicht möglich.

Friedel Grünthal

Christian Dutz

Bernadette Linasi

Bernadette Linasi and Christian Dutz:

About 12 miles from Gut Aiderbichl in Henndorf, phones are being answered, countless clicks are sent with their online missions, and reservations are made and sold out while Holde Sudenn takes reservations from travel groups and/or arranges live footage events with TV crews. Then there's Friedel Grünthal who keeps us up-to-date by faxing us animal reports from around the globe.

There are times when the lights stay on deep into the night, like when I'm sitting with Christian in front of a computer I don't even know how to use. It's one of a number of heartbeats that keep Gut Aiderbichl alive—it may be in the background, but it's just as strong as those of the animals outside. Without a control center, none of this would be possible at all.

Dr. Hergard Spielvogel

Anna Pieringer

Holde Sudenn

Tierschutz ist zugleich Menschenschutz
Wie Aiderbichl entstand

Ich, Michael Aufhauser, wurde am 25.04.1952 in Augsburg geboren. Nach Abschluss einer dreijährigen Schauspielschule und einigen Jahren als Schauspieler wechselte ich in die amerikanische Tourismusindustrie. Über 20 Jahre lang lebte ich auf verschiedenen Kontinenten und wurde Vize-Präsident eines weltweit agierenden Unternehmens in Boston.

„Perrera", Málaga

Unter anderem gab es auch ein Büro an der Costa del Sol in Spanien. Dort beobachtete ich vor über 20 Jahren, wie ein Straßenhund, den ich schon länger kannte, von Hundefängern weggeschafft wurde. Ich beschloss, deren Fahrzeug zu folgen und wurde kurz darauf Zeuge der grauenvollen Vergasung von ca. 40 Hunden und Katzen in der „Perrera" in Málaga. Die verbliebenen Tiere, es waren weit über 30, kaufte ich frei und sorgte für deren Unterbringung und Vermittlung. Es war, als hätte ein Blitz in mein Leben eingeschlagen. Nichts blieb, wie es vorher war. Immer wieder kehrte ich zu der Gaskammer zurück und verhalf vielen hundert Hunden zur Freiheit.

Schon zu dieser Zeit schmiedete ich mit meiner langjährigen Lebensbegleiterin Irene Florence Pläne, eines Tages eine Begegnungsstätte zwischen Mensch und Tier zu errichten.

Im Jahr 1999 war es dann soweit. Ein wunderschöner Ort wurde gefunden – ganz nahe bei Salzburg, in Henndorf am Wallersee. Anfangs war lediglich die Unterbringung von 25 Pferden, unter anderem aus privatem Besitz, geplant, aber bald stellte sich heraus, dass dies der Ort werden sollte, von dem Irene Florence und ich vor langer Zeit geträumt hatten. 2001 fand ein ökumenischer Tierschutz-Gottesdienst auf Gut Aiderbichl statt. Dass dieser Tag eine Art Einweihung werden sollte, nicht nur für einen Gnadenhof, sondern für eine Bewegung der ganz besonderen Art, ahnte damals noch niemand.

Tierschutz ist zugleich Menschenschutz. Das hatte so noch niemand vorher formuliert. Dennoch liegt es auf der Hand. Solange wir Tiere vor uns Menschen schützen müssen, haben wir ein grundsätzliches Problem. Denn unser Umgang mit Tieren spiegelt unseren Umgang miteinander und insbesondere mit Schwächeren wider.

Zuerst galt es jedoch, das Credo, dass Tiere keine Sachen sind, in den Medien dauerhaft zu positionieren. Für die beispiellose Medienarbeit von Gut Aiderbichl spielte meine Marketingerfahrung eine große Rolle. In Amerika hatte ich die Befähigung erworben, ein junges Team zum Mitmachen zu motivieren und an der Entwicklung einer Idee teilhaben zu lassen. Diese Erfahrung brachte ich bei meiner neuen Tätigkeit ein, was Gut Aiderbichl zu dem Stellenwert verhalf, den es in der Gesellschaft heute überregional hat.

Protect Animals and You Protect Humans
how Aiderbichl Began

My name is Michael Aufhauser. I was born on April 25, 1952, in Augsburg, Germany. After graduating from a three-year drama school program, I'd spent several years as an actor and then decided to work in tourism in the United States. After spending more than 20 years in various continents, I became the vice president of a globally active company in Boston.

My position included having an office in the Costa del Sol region in Spain. It was there, more than 20 years ago, when I saw some dogcatchers taking away a stray dog I had come to know. Deciding to follow them, I then had to witness a scene in which approximately 40 dogs and cats were horribly gassed to death at the "Perrera", Málaga. I immediately bought the surviving animals—more than 30 of them—and made sure they were placed in proper homes. It altered my life in a very profound way. Nothing was ever the same again. Time and time again, I kept returning to that gas chamber and gave many hundreds of more animals their freedom.

During that time, along with my lifelong partner, Irene Florence, I was already working on plans to establish a future place where humans could enjoy encounters with animals.

In 1999, our plans became a reality. A beautiful place was found very close to Salzburg, Austria, by the town of Henndorf at Lake Wallersee. Our initial plans only called for keeping 25 horses, some of which we owned privately. However, it wasn't long before Irene Florence and I realized that we'd final-ly found the place we'd been looking for all those years. In 2001, Gut Aiderbichl was ecumenically dedicated to the protection of animals. Nobody who was there could have foreseen that that day would not only mark the dedication of an animal sanctuary, but also a very special movement.

Protect animals and you protect humans. Nobody ever put it like that, right? But here's the thing: As long as we have to protect animals from ourselves, we're always going to have a basic problem. You see, the way we treat animals reflects the way we treat each other and especially those who can't defend themselves.

The first thing we had to do, however, was to solidify in the media our article of faith that animals are not objects. The extraordinary media campaign of Gut Aiderbichl to this end has largely been the result of my marketing experience. During my stay in America, I had acquired the ability to motivate a young team to become involved in an idea and to help develop it. By injecting this ability into my new activity I helped Gut Aiderbichl obtain the nationwide significance in society it enjoys today.

Besuch im Altenheim
A visit to a seniors' home

Das Hauptaugenmerk liegt auf der Vermittlung von Werten, die uns weitgehend verloren gegangen sind, und auch darauf, für das Thema „Tiere" eine breite Öffentlichkeit zu mobilisieren. Messbarer als die Verbreitung unserer Philosophie ist die Arbeit im Bereich der Gnadenhöfe. Unter dem Schutz von Gut Aiderbichl stehen derzeit auf elf Höfen über 1000 gerettete Tiere in bester Haltung. Die beiden größten, in Henndorf bei Salzburg und in Eichberg bei Deggendorf, können an 365 Tagen im Jahr besucht werden. Was Besucher dort vorfinden, überzeugt und hat viele zu Mitstreitern gemacht. Die Biografien der aus großen Notlagen geretteten Tiere dokumentieren, wie wir Menschen mit ihnen umgegangen sind und was wir verbessern müssen, wenn wir kulturfähig bleiben wollen.

Gut Aiderbichl hat eine Reihe von Erfolgen erzielt, die uns gleichermaßen Stolz machen und Ansporn für weiteres Engagement liefern: Wir haben Freilaufparks für Hunde geschaffen, maßgeblich dazu beigetragen, dass Tierschutz im Land Salzburg Verfassungsrang erhalten hat, Begegnungen geschaffen mit behinderten Menschen, Hospizbewohnern und Senioren, denen der Kontakt zu Tieren etwas Licht und Freude in ihre schwere Lebenslage bringt.

In our activities, we focus on promoting the kind of values that many of us in society seem to have lost and also on mobilizing the public at large when it comes to the situation of animals. But promoting our philosophy doesn't produce the same results as our work at the sanctuaries does. Gut Aiderbichl currently has eleven sanctuaries giving the best care to more than 1,000 saved animals. The two largest of these, at Henndorf near Salzburg, Austria, and at Eichberg near Deggendorf, Germany, are open to visitors 365 days a year. What our visitors find there is convincing enough to have made many of them active supporters of our cause. The biographies of many animals saved from dire situations exemplify the way we humans have treated them and what we have to do better, if our future as a civilized society means anything to us at all.

Gut Aiderbichl has enjoyed a number of successes that make us proud as much as eager to push forward in our cause. We've created off-leash areas for dogs, we've been a major driving force behind adding the protection of animals to the regional constitution of Salzburg Province, and we've organized encounters between animals and handicapped people, hospice patients and the elderly to make their difficult situations a little brighter and friendlier.

PATENSCHAFTEN

Zur Philosophie von Gut Aiderbichl: Es waren die Menschen, die sich für Gut Aiderbichl engagieren, die einen Begriff geprägt haben, der in drei Worten ihre Gesinnung beschreibt: „Ich bin Aiderbichler!" Mit Gut Aiderbichl haben sie einen Ort und Menschen gefunden, die ihre Art von Tierliebe, ihre Gedanken, Gefühle und Hoffnungen verstehen. Und deshalb haben sie sich entschlossen, „Aiderbichler" zu werden. Über 1000 Tiere stehen derzeit auf elf Höfen in Deutschland und Österreich unter unserem Schutz.

Wie kann ich „Aiderbichler" werden und die Anliegen von Gut Aiderbichl unterstützen?

Pate/Mitglied werden: Zum Beispiel mit einer Paten-/Mitgliedschaft, die schon ab €10,00 monatlich möglich ist und Ihnen und Ihren Begleitpersonen an 365 Tagen im Jahr nicht nur freien Eintritt auf unseren Gütern in Henndorf und Deggendorf bietet, sondern auch Zugriff auf unsere Live-Kameras im Internet und vieles mehr.

Tierpatenschaften: Gut Aiderbichl unterhält Gnadenhöfe der besonderen Art und ist kein klassisches Tierheim. Die Haltung unserer Tiere ist auf einen lebenslangen Verbleib bei uns ausgerichtet. Deshalb verfügen wir über einen Stamm gut ausgebildeter Tierpfleger, beste medizinische Versorgung und eine Haltung jenseits von Zwingern und Käfigen. Außerdem gehen wir auf die Individualität der Tiere, zum Beispiel ihre Freundschaften und ihren Rang innerhalb der Gruppe besonders ein. Sie leben bei uns ohne Druck und Angst. Patenschaften können auf Wunsch mit der Benennung eines einzelnen Tieres, das bei uns lebt, abgeschlossen werden. Die Einnahmen kommen allen geretteten Tieren zugute.

Förderer der Gut Aiderbichl Stiftung werden: Außer den beiden großen bekannten Gütern gibt es noch neun weitere Höfe, auf denen viele von uns gerettete Tiere leben. Sie werden ausschließlich von den gemeinnützigen Gut Aiderbichl Stiftungen in Deutschland und Österreich unterstützt. Bitte helfen Sie uns. Spenden Sie und werden Sie schon mit einer einmaligen Spende Förderer von Gut Aiderbichl.

Um „Aiderbichler" zu werden, wenden Sie sich bitte an: Gut Aiderbichl Verwaltung, Johannes-Filzer-Straße 5, 5020 Salzburg oder telefonisch an: +43 (662) 62 53 95 oder per E-Mail an: info@gut-aiderbichl.com.

Jährlicher Weihnachtsmarkt auf Gut Aiderbichl / Annual christmas fair at Gut Aiderbichl

SPONSORSHIPS

About the Philosophy of Gut Aiderbichl: It was the people dedicating themselves to Gut Aiderbichl, who coined a phrase that puts their philosophy into three words, "I am an Aiderbichler!" In Gut Aiderbichl, they have found a place and people that identify with their level of love for animals, their thoughts, feelings, and hopes. That's what made them decide to become an "Aiderbichler." Currently, we have more than 1,000 animals under our care and protection in eleven sanctuaries in Germany and Austria.

How Can I Become an "Aiderbichler" and Support the Cause of Gut Aiderbichl?

Becoming a Sponsor/Member: A sponsorship or membership, for example, is available starting at just €10.00 ($15) a month. Not only does it provide you and traveling companions free admission to our sanctuaries in Henndorf and Deggendorf, but also with access to our live cameras on the Internet and a whole lot more.

Animal Sponsorships: Gut Aiderbichl operates a unique group of animal sanctuaries. It's not a protection shelter in the classical sense. Our care and provision is geared towards the idea of our animals spending the rest of their lives with us. That's why we have a team of highly qualified dedicated caretakers, first-rate medical supplies, and a care system that never relies on kennels or cages. We also practice a keen awareness of our animals' individual characters, including their relationships and ranks among their groups. We give them a life free of any stress or fear. On request, sponsorships can be arranged just by naming an individual animal that lives with us. Proceeds are used for the benefit of all rescued animals.

Becoming a Promoter of the Gut Aiderbichl Foundation: In addition to our two major and well-known sanctuaries, we have nine more sanctuaries sheltering many animals we've rescued. They are exclusively supported by our non-profit Gut Aiderbichl Foundations in Germany and Austria. Please help us. By making just one donation, you can become a promoter of Gut Aiderbichl.

To become an "Aiderbichler," please contact: Gut Aiderbichl Verwaltung, Johannes-Filzer-Straße 5, 5020 Salzburg, Austria, phone +43 (662) 62 53 95, or send your e-mail to: info@gut-aiderbichl.com.

Europas größte Lebendtierkrippe / Europe's greatest nativity crib with live animals

IMPRINT

© 2008 teNeues Verlag GmbH + Co. KG, Kempen
Text and photographs © Gut Aiderbichl GmbH
All rights reserved.

Editor: Michael Aufhauser
Gut Aiderbichl Stiftung, Gut Aiderbichl GmbH, Johannes-Filzer-Str. 5, A-5020 Salzburg
Responsible for content: Michael Aufhauser, Dieter Ehrengruber, Friederike Grünthal
Associates: Christian Dutz, Michaela Kalss, Sabine Schlömer, Helmut Schödel, Holde Sudenn
Photographs by Dieter Ehrengruber, Andreas Kolarik, Franz-Josef Lang, Franz Neumayr, Agnes Schindler,
Alexandra Schlump, Markus Tschepp, Jürgen Weyrich, Zeppelzauer
Translation in association with Artes Translations: Conan Kirkpatrick
Design by Eva Reuters
Production by Sandra Jansen
Editorial coordination by Pit Pauen
Color separation by MT-Vreden, Vreden

Published by teNeues Publishing Group

teNeues Verlag GmbH + Co. KG
Am Selder 37
47906 Kempen, Germany
Tel.: 0049-(0)2152-916-0
Fax: 0049-(0)2152-916-111
e-mail: books@teneues.de

Press department: Andrea Rehn
Tel.: 0049-(0)2152-916-202
e-mail: arehn@teneues.de

teNeues Publishing Company
16 West 22nd Street
New York, NY 10010, USA
Tel.: 001-212-627-9090
Fax: 001-212-627-9511

teNeues Publishing UK Ltd.
P.O. Box 402
West Byfleet
KT14 7ZF, Great Britain
Tel.: 0044-1932-4035-09
Fax: 0044-1932-4035-14

teNeues France S.A.R.L.
93, rue Bannier
45000 Orléans, France
Tel.: 0033-2-3854-1071
Fax: 0033-2-3862-5340

www.teneues.com

ISBN: 978-3-8327-9276-3

Printed in Italy

Picture and text rights reserved for all countries.
No part of this publication may be reproduced in any manner whatsoever.
All rights reserved.
While we strive for utmost precision in every detail, we cannot be held responsible
for any inaccuracies, neither for any subsequent loss or damage arising.
Bibliographic information published by the Deutsche Nationalbibliothek.
The Deutsche Nationalbibliothek lists this publication in the Deutsche
Nationalbibliografie; detailed bibliographic data are available in the
Internet at http://dnb.d-nb.de.

teNeues Publishing Group
Kempen
Düsseldorf
Hamburg
London
Munich
New York
Paris

teNeues